Digital Communications

Design for the Real World

Andy Bateman

Addison-Wesley

Harlow, England ● Reading, Massachusetts ● Menlo Park, California

New York ● Don Mills, Ontario ● Amsterdam ● Bonn ● Sydney ● Singapore

Tokyo ● Madrid ● San Juan ● Milan ● Mexico City ● Seoul ● Taipei

Addison Wesley Longman Limited
Edinburgh Gate
Harlow
Essex CM20 2JE
England

and Associated Companies throughout the World.

Cover incorporating photograph © PhotoDisc
Typeset in Stone Sans and Serif by 43
Produced by Addison Wesley Longman Singapore (Pte) Ltd., Printed in Singapore

First printed 1998

ISBN 0-201-34301-0

British Library Cataloguing-in-Publication Data
A catalogue record for this book is available from the British Library

Contents

6 Multi-level digital modulation 141

8 Multi-user digital modulation techniques

Preface

Who is this book for?

This book provides a framework for understanding and evaluating the key design topics and choices involved in developing a data communications link.

It has been devised specifically to serve as a *first stage text* for undergraduate students (probably in their second year of study), such that they are fully equipped to study the more theoretical, statistical aspects of digital communications in later years, and will do so fully conversant with the context in which this analysis and rigour will be applied.

By bringing to the fore the design choices facing communications engineers, within an intuitive framework, the book also fulfils the role of a *Getting Started* plus *Frequently Asked Questions* manual in digital communications for practising engineers and managers. It will allow them to home in rapidly on the design parameters that are pertinent to their particular application and specification, and so better focus engineering resource within their organizations.

What previous knowledge is required?

Readers of the book are expected to have a basic understanding of the frequency content of simple waveforms (Fourier Series expansions) and hence appreciate the concepts of bandwidth, filtering and so on.

In addition, the mathematics of integration, differentiation and trigonometry are assumed, but not relied on for understanding of the text. For those who are a bit rusty in these areas, a short review of Fourier series and relevant trigonometrical relationships is included in Chapter 1.

All other material is developed from an intuitive or first principles approach.

How does the material in this book differ from other digital communications texts?

Most, if not all, competing student texts aim to cover the full subject matter over all years of a degree programme. As a consequence, the subject matter is predominately driven from a mathematical description of modulation type and performance, in contrast to the application and intuitive approach of this text.

The subject material in this book is a carefully selected range of topics that give the reader a structured overview of the digital communications arena, with emphasis on modem design, performance trade-offs, key limiting factors, and practical issues in implementation over real-world channels.

The book provides a *layered framework* in which the reader can quickly identify those design aspects relevant to his or her application and then follow the hyperlinks, or reference list provided, in order to obtain more specific information.

Where has the material originated?

The material within the book has been developed and refined over ten years of lecturing on digital communications to second and third year undergraduates and Masters level students at Bristol University in England, in a series of focused courses to engineers and managers run by Oxford University, and on site with companies such as Hewlett-Packard and Philips.

Experience has repeatedly shown that practising engineers who have pursued an undergraduate or post-graduate communications course in the past welcome the applications-driven and context-driven structure of this book, serving as both a reminder of the key design issues and a pooling of the basic design information and criteria from which to structure a more detailed research or design programme.

About the author

Andy Bateman was formerly a Professor of Communications and Signal Processing at the University of Bristol, England. Aged 39, he has been involved in the research and design of digital communications systems for the past 17 years, working with companies such as Motorola, Nokia, Ericsson, AT&T and Hewlett-Packard to develop state-of-the-art data communications equipment. In 1995 he co-founded a company – Wireless Systems International Ltd (www.WSIL.com) – which specializes in providing equipment and design/consultancy in all aspects of wireless digital communications, with clients ranging from small and large telecommunications manufacturers to operators and end users. He can be contacted on email at: ab@wsi.co.uk

As a regular invited lecturer on the subject of digital communications design around the world, and with the challenging task of introducing second year undergraduates at Bristol to the vast topic of modern communications, he has developed a very intuitive means of presenting the often complex material so that the principles are easily grasped. This style makes the book very readable as both an excellent basic reference and introductory student text.

He is also author of the book *Digital Signal Processing Design* (Bateman and Yates (1989)), which is a practical guide to DSP algorithms and applications.

Information for instructors/trainers

Files containing the figures from this book in Portable Document (pdf) format for use as transparencies will be made available for adopting lecturers or trainers. Please contact your local Addison Wesley Longman representative for information.

If you intend to network this product within your institution or company, please contact the following for further information:

Interactive Learning Europe	Tel	+44 (0)1223 425558 x 787
124 Cambridge Science Park	Fax	+44 (0)1223 425349
Milton Road	Email	sean.massey@awl.co.uk
Cambridge CB4 4ZS	Web	http://www.awl-ile.com

Interactive Learning Europe specializes in software to facilitate interactive learning in Higher and Further Education, with an emphasis on software that can be used both professionally and by students.

What do you think about this book?

If you have any feedback on either the book or CD please contact us at:

engineering.feedback@awl.co.uk

or visit the book's homepage at:

http://www.awl-he.com/engineering/bateman.html.

Alternatively, click on the **Feedback** button from any page within the electronic version.

Acknowledgements

I would like to express my thanks to Anna Faherty and Dylan Reisenberger at Addison Wesley Longman for their enthusiasm and professional approach in the preparation of this book, and Mike Smith for his magic with the JavaScript.

I would also like to thank my two sons, Callum (8) and Jamie (6), for letting me have time on my computer, and my wife Jacqui ( ) for her encouragement and patience throughout and for her help with the composition of the CD's title music. This book is dedicated to them and to my wider family spread throughout the world.

Publisher's Acknowledgements

The publishers are grateful to the following for permission to reproduce material featured in the book and on the CD: to Professor David Bull, Centre for Communications Research, University of Bristol, for images showing intelligent source coding on book page 172; to Motorola Limited for Semiconductor Technical Data sheets featured in in-depth sections on book pages 110, 120 and 170; to Wireless Systems International Limited for images of the output of a cartesian loop amplifier with TETRA modulation input, the implementation of a CALLUM transmitter, the image of the feedforward amplifier and the image of its output, featured on book pages 80–83; and for the plot on book page 101, top. We are grateful to Corel for permission to feature certain clipart in artwork in this product. The images in the electronic product are only to be used for viewing purposes and may not be resaved or redistributed.

Trademark Notice

Internet Explorer and Windows are trademarks of Microsoft Corporation. Netscape, Netscape Navigator, Netscape Communicator and the Netscape logos are trademarks of Netscape Communications Corporation. Netscape Communications Corporation has not authorized, sponsored, endorsed or approved this publication and is not responsible for its content.
Macintosh is a trademark of Apple Computer, Inc.
Matlab is a trademark of The Mathworks, Inc.
OAK DSP core is a registered trademark of LSI Logic Corporation.
TMS 320C6X/TMS320C6201 series are trademarks of Texas Instruments.
DCS1800 is a trademark of Digital Communications.

How to use the combined book and CD

Both the printed book and electronic book contain the same fundamental content on a page-by-page basis. They can therefore be used independently and interchangeably, according to the preferred learning style of the student, teaching style of the department, or method of easiest reference for the professional.

Differences between the printed book and the CD

You may read the printed book without reference to the CD as the book contains all the fundamental information you need to learn about digital communications. The CD contains the same content as the book in browsable electronic form along with these additional features to enhance the learning experience:

- animated figures to aid understanding

- linked cross-references for ease of exploration

- hyperlinks to further information on the World Wide Web

- Matlab code, ready to run simulations

- answers to problems from the printed text.

In the book, icons indicate where the animations, weblinks and Matlab code feature on the CD.

indicates that the figure is animated on the CD.

indicates that the CD has a link to information on the Web about the topic underlined.

indicates that the Matlab code generating that graph/plot is available on the CD.

Systems requirements

The software on the CD is in 16-bit form and is designed for use with PC/Windows (3.1, '95, NT) and Macintosh machines with at least 8 Mb RAM. It has been optimized for use on Pentium PCs or PowerMACs.

Hyperlinks from the product to related material on the World Wide Web will only function if you have an active Internet connection already set up.

The product supports its own bookmarking function using the browser's internal Cookie software. If you usually disable Cookies in your browser you will be unable to bookmark pages of the electronic book.

The viewing window has been designed for optimum layout of the text and graphics without need for scrolling. However, at screen resolutions below the common SuperVGA (800×600) you may have to scroll to access all the information on the screen.

The electronic book is designed to be viewed using Microsoft Internet Explorer version 4.0 or later. Some features may not operate in other browsers.

Installing and starting the electronic book

The book files are designed to be viewed using the Microsoft Internet Explorer browser version 4.0 or later. To install the browser on your machine, see notes in the README on the CD.

To start the 'book' after installing the browser:

1. Start Internet Explorer 4.

2. (PC) Choose **File** from the Menubar and **Open** from the File menu. Click on the **Browse** button to obtain a file listing.

2. (MAC) Choose **File** from the Menubar and **Open file...** from the File menu.

3. Choose the drive and directory/folder containing the product (eg CD drive or network drive if the product is networked), and open the file **start.htm**. This launches the opening screen. You may want to add this page to your browser's 'favourites' so that you can locate it quickly next time.

4. Click on the **Start** button. The main window for the electronic book will appear. If not all of the window is visible, reposition the window and/or see the notes under systems requirements above.

Navigating around the electronic book

Buttons on the left-hand side of the screen

You can navigate to any part of the book by using the buttons appearing down the left-hand side of the screen.

How to use/ Preface Click on **How to use/Preface** to get instructions on how to use the electronic product or to read the Preface of the book.

Contents Click on **Contents** to see the contents of the entire book. Clicking on any topic in the contents will link directly to the relevant page.

Chapter buttons eg **1 Background material** To start exploring a new chapter, click on the relevant chapter button, eg **1 Background material**. You will see a list of sections available in that chapter. Choose a section to explore, either by clicking on a topic in the list or by using the section tabs (see below). Alternatively, you may go back to the page you were looking at before by using the **Back** button (see over).

Section tabs Click on a **section tab** to see the contents of that section within the current chapter. You will see a list of pages available within that section. Choose a page to begin reading at by clicking on a topic in the list. Alternatively, you may go back to the page you were looking at before by using the **Back** button (see over).

At any time, the chapter and section you are reading are highlighted by a change in colour of the chapter buttons and section tabs. The chapter and section number and title of the current section are also given in small print at the top of the main page.

Glossary Click on **Glossary** to see the glossary of abbreviations and other terms.

References Click on **References** to see the list of references for the book.

Index Click on **Index** to see an index to the electronic book. Clicking on a topic within the index will link you back to the key entry for that subject in the book.

0.2 Preface

Preface

Who is this book for?

This book provides a framework for understanding and evaluating the key design topics and choices involved in developing a data communications link.

It has been devised specifically to serve as a *first stage text* for undergraduate students (probably in their second year of study), such that they are fully equipped to study the more theoretical, statistical aspects of digital communications in later years, and will do so fully conversant with the context in which this analysis and rigour will be applied.

By bringing to the fore the design choices facing communications engineers, within an intuitive framework, the book also fulfils the role of a *Getting Started* plus *Frequently Asked Questions* manual in digital communications for practising engineers and managers. It will allow them to home in rapidly on the design parameters that are pertinent to their particular application and specification, and so better focus engineering resource within their organizations.

What previous knowledge is required?

Readers of the book are expected to have a basic understanding of the frequency content of simple waveforms (Fourier Series expansions) and hence appreciate the concepts of *bandwidth*, *filtering* and so on.

In addition, the mathematics of *integration*, *differentiation* and *trigonometry* are assumed, but not relied on for understanding of the text. For those who are a bit rusty in these areas, a short review of *Fourier series* and relevant *trigonometrical relationships* is included in chapter 1.

All other material is developed from an intuitive or first principles approach.

Next page
Previous page
Back
Bookmark this page
View bookmarks
Feed-back

Buttons at the foot of the screen

Further buttons at the foot of the screen aid navigation and provide links to AWL's website.

Updates/ Feedback	Click on **Updates/Feedback** to jump to the book's homepage on the AWL website (`http://www.awl-he.com/engineering/ bateman.html`), where you can send us feedback and find updates and supplementary material for the book. (This will only function if you have an active Internet connection already set up.)
Bookmark this page	Click **Bookmark this page** to add the page you are looking at to your list of bookmarks. You can bookmark up to twenty individual pages.
View bookmarks	Click **View bookmarks** to see a list of all pages you have bookmarked. You may link directly to any bookmarked page by clicking on the underlined title. You may delete any existing bookmark by clicking on its **Delete** button.
Back	Click on **Back** to go back to the page you were reading before the current one (not necessarily the same as the page that precedes the current one). The **Back** button will work more than once in sequence, eg you can go back to the page you were looking at three before the current one by clicking back three times.
Previous	Click on **Previous** to jump to the page that precedes the current page in the book's sequence.
Next	Click on **Next** to jump to the page that follows the current page in the book's sequence.

Features within the page and display conventions

The following display conventions and features can be found within individual pages.

Pale blue text	is used to highlight important terms and concepts.
Blue underlined text	indicates a cross-reference to further information within the book. Clicking on the link takes you to the topic, so for example you can follow a link to learn about a new, related topic or to refresh your memory about an old one. This linked approach encourages flexible learning styles and allows the book to be layered, with more in-depth information, or more detailed explanation/revision, only offered to you if requested or required.

<u>**Red underlined text**</u>	indicates a hypertext link to related information on the World Wide Web. Clicking on the link opens a new window containing the relevant page. These links will only function if you have an active Internet connection already set up.
Blue panel	is used to indicate definitions of key concepts.

Many pages are accompanied by further material in the form of in-depth sections, worked examples, questions and Matlab code. Where there is further material associated with a main page, relevant buttons appear in the top-right corner of the main page. Clicking on the button will open a new window which contains the requested information and can be viewed simultaneously with the main book screen. This linked approach encourages flexible learning styles and allows the book to be layered, with more in-depth information, or more detailed explanation/revision, only offered to you if requested or required.

In-depth

The **in-depth** link in the top-right corner of a main page indicates that more detailed text relating to this topic is available. This might be more detail of the mathematics behind the material introduced on the page, or other reference material. Clicking on the link launches a separate window containing the in-depth material. You may leave the in-depth window open and continue browsing in the main window, if you want to refer to the in-depth again later; otherwise use the **Close** button to close it.

Example #.#

The **Example #.#** link in the top-right corner of a main page indicates that there is a worked example relating to this material. Clicking on the link launches a separate window containing the relevant example. Whilst in the example window you may also access other examples within that chapter by using the dropdown menu. You may leave the example window open and continue browsing in the main window, if you want to refer to the example again later; otherwise use the **Close** button to close it.

Question #.#

The **Question #.#** link in the top-right corner of a main page indicates that there is a question relating to this material. Clicking on the link launches a separate window containing the relevant question. The **Show answer** button reveals the solution to the question. Whilst in the question window you may also access other questions within that chapter by using the drop-down menu. You may leave the question window open and continue browsing in the main window, if you want to refer to the question again later; otherwise use the **Close** button to close it.

Matlab

MAT LAB

The **Matlab** link in the top-right corner of a main page indicates that the Matlab code which generated the graph/plot on the page is available on the CD. You can use the code with Matlab version 4 onwards to resimulate the results and you may be able to vary relevant parameters. Clicking on the link will tell you the name of the relevant Matlab file on the CD, and what it will allow you to do.

1 Background material

Before we can make sense of the issues important in the design of a digital communications link, it is essential to have a good grounding in the relationship between the shape of a digital waveform in the *time domain*, and its corresponding spectral content in the *frequency domain*.

Many of the modulation processes to be described in this book become intuitive when working with the *sine* and *cosine* terms that make up the modulating signals. In contrast, in the time domain we have to deal with the more complex issues of correlation and convolution. To this end, the first two sections of this chapter present the basics of Fourier series and trigonometrical relationships, as a starting point or refresher course for the reader as required.

The chapter also provides a broad-brush overview of *network and protocol* aspects of data communications. This brief summary cannot do justice to the subject, which in some respects commands the lion's share of a digital communications system. The focus of this book however is on the modem part of the link, where we are concerned with getting the data bits (1s and 0s) which appear at the network interface over the channel in the most efficient (as regards cost, power, bandwidth, time) and error free manner. It is here, at the physical interface layer, that many of the biggest design challenges occur.

The final section of the chapter provides a summary of terms and ideas used throughout the book, so that jargon does not defeat the reader at the first hurdle.

1.1 Time/frequency representation of digital signals

It will become apparent after reading the early chapters of this book that the performance of a digital communications link is constrained by two primary factors: channel bandwidth and system noise.

In order to understand fully the interaction between system data/symbol rate, modulation type, pulse shape and channel bandwidth, a grasp of the frequency content of various types of time domain data signals is key.

The mathematical tools used to map between the time and frequency domain are most commonly the Fourier series representation (for periodic signals) and the Fourier transform (for general periodic and non-periodic signals). Other mapping techniques exist, such as wavelets, but are beyond the scope of this book.

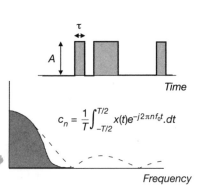

$$c_n = \frac{1}{T}\int_{-T/2}^{T/2} x(t)e^{-j2\pi n f_0 t}.dt$$

Fourier series

Fourier series can represent any periodic time domain signal by a summation of harmonically related sinewaves.

For example, the square wave (equivalent 1,0,1,0,1,0,...) data signal shown here can be constructed from sinewaves of descending amplitudes, spaced, in this example, at odd multiples of the *fundamental frequency* of the square wave.

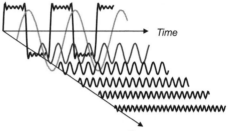

If we wished to represent the 1,0,1,0,1,0,... pattern perfectly, an infinite number of sinusoidal components would be required, implying that we need an infinite channel bandwidth!

If, as is always the case in practice, the channel has a finite bandwidth, then we can expect to receive a non-perfect replica of the input time domain waveform, unless all of its Fourier series components fall within the available channel bandwidth.

The output response of a channel passing only the first three frequency components of the square wave is shown here and clearly demonstrates the change caused by restricted bandwidth on the time domain waveform. This example also shows, however, that the 1,0,1,0,1,0,... data stream can still easily be detected without all the constituent frequency components of the original square wave, and in fact correct demodulation is possible if only the *fundamental component* is passed by the channel.

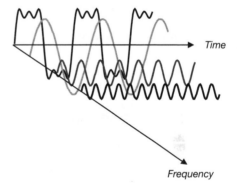

IN DEPTH

Fourier series expansion

The Fourier series representation of a time domain waveform is usually written as a trigonometric or exponential expansion, taking the following form:

Trigonometric expansion

$$x(t) = a_0 + \sum_{n=1}^{\infty} (a_n \cos(2\pi n f_0 t) + b_n \sin(2\pi n f_0 t))$$

where

$$a_0 = \frac{1}{T} \int_0^T x(t) = \text{average signal energy}$$

$$a_n = \frac{2}{T} \int_0^T x(t) \cos(2\pi n f_0) \cdot dt$$

$$b_n = \frac{2}{T} \int_0^T x(t) \sin(2\pi n f_0 t) \cdot dt$$

Complex exponential expansion

$$x(t) = \sum_{n=-\infty}^{\infty} c_n e^{j2\pi n f_0 t}$$

where

$$c_n = \frac{1}{T} \int_{-T/2}^{T/2} x(t) e^{-j2\pi n f_0 t}$$

The frequency domain

The representation of a time domain signal by a summation of sine or cosine components is usually referred to as the *spectrum* of the waveform. It is traditional to draw the spectrum as discrete lines on a graph, with the position of the lines on the *x*-axis representing the frequency of the component, and the height of the line representing the amplitude.

Again, it is common practice to represent only the absolute value or magnitude of each component in the spectrum; however, one must not forget that in fact each term in the Fourier series expansion could be made up of both sine and cosine terms at any given frequency and hence both *magnitude* and *phase* are required for a complete representation of the time domain signal.

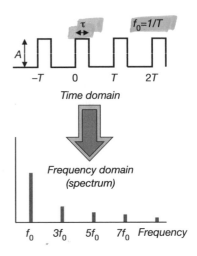

Time domain

Frequency domain (spectrum)

Spectrum of a periodic pulse train

If we now consider the Fourier series expansion of a train of pulses representing successive data bits, we find (see examples) that the amplitudes of the frequency components are all constrained by a general *spectral envelope* which passes through zero at multiples of the data pulse width τ.

This spectral envelope is given by the equation:

$$\text{sinc envelope} = \frac{2A\tau}{T} \cdot \frac{\sin(\pi n \tau/T)}{(\pi n \tau/T)}$$

which is usually termed the *sinc* function.

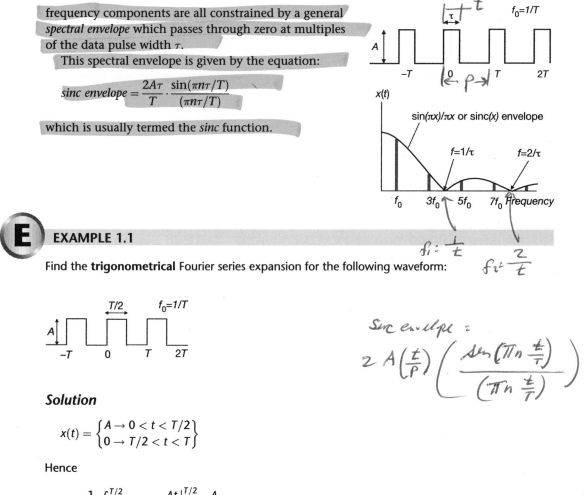

E **EXAMPLE 1.1**

$$f_i = \frac{1}{t}$$

Find the **trigonometrical** Fourier series expansion for the following waveform:

$$f_{i^2} = \frac{2}{t}$$

Sinc envelope :

$$2A\left(\frac{t}{P}\right)\left(\frac{\sin\left(\pi n \frac{t}{T}\right)}{\left(\pi n \frac{t}{T}\right)}\right)$$

Solution

$$x(t) = \begin{cases} A \to 0 < t < T/2 \\ 0 \to T/2 < t < T \end{cases}$$

Hence

$$a_0 = \frac{1}{T}\int_0^{T/2} A \cdot dt = \frac{At}{T}\bigg|_0^{T/2} = \frac{A}{T}(T/2)$$

$$\Rightarrow a_0 = A/2$$

$$a_n = \frac{2}{T}\int_0^{T/2} A \cdot \cos 2\pi n f_0 t \cdot dt = \frac{2A}{2\pi n f_0 T} \cdot \sin 2\pi n f_0 T/2$$

Now,

$$f_0 = 1/T$$

$$\Rightarrow \therefore a_n = \frac{A}{\pi n}\sin \pi n = 0 \text{ for all } n$$

$$b_n = \frac{2}{T}\int_0^{T/2} A \sin 2\pi n f_0 t \cdot dt = \frac{-2A}{2\pi n f_0 T} \cdot (\cos(2\pi n f_0 T/2) - \cos 0)$$

$$\Rightarrow b_n = \frac{A}{\pi n}(1 - \cos \pi n) = \frac{2A}{\pi n} \text{ for } n \text{ odd and 0 for } n \text{ even}$$

The full Fourier series expansion can thus be written as:

$$\Rightarrow x(t) = A/2 + \sum_{n=1,3,5,\ldots}^{\infty} \frac{2A}{\pi n} \cdot \sin 2\pi n f_0 t$$

If the waveform is shifted by $T/4$, the Fourier series expansion would be represented by cosine terms as follows:

$$\Rightarrow x(t) = A/2 + \sum_{n=1,3,5,\ldots}^{\infty} \frac{2A}{\pi n} \cdot \cos 2\pi n f_0 t$$

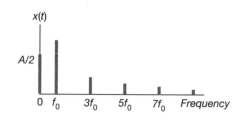

E EXAMPLE 1.2

Find the **trigonometrical** Fourier series expansion for the following waveform:

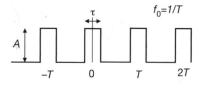

Solution

The general expression for the trigonometrical Fourier series expansion of a function $x(t)$ is given by

$$x(t) = a_0 + \sum_{n=1}^{\infty}(a_n \cos 2\pi n f_0 t + b_n \sin 2\pi n f_0 t)$$

where

$$a_0 = \frac{1}{T}\int_0^T x(t) \cdot dt$$

$$a_n = \frac{2}{T}\int_0^T x(t) \cos(2\pi n f_0 t) \cdot dt$$

$$b_n = \frac{2}{T}\int_0^T x(t) \sin(2\pi n f_0 t) \cdot dt$$

For the waveform shown above:

$$x(t) = \begin{cases} A \rightarrow -\tau/2 < t < \tau/2 \\ 0 \rightarrow \tau/2 < t < T - \tau/2 \end{cases}$$

Hence

$$a_0 = \frac{1}{T} \int_{-\tau/2}^{\tau/2} A \cdot dt = \frac{A}{T} \cdot t \Big|_{-\tau/2}^{\tau/2} = \frac{A\tau}{T}$$

$$\Rightarrow a_0 = A\tau/T$$

$$a_n = \frac{2}{T} \int_{-\tau/2}^{\tau/2} A \cos 2\pi n f_0 t \cdot dt = \frac{2A}{2\pi n f_0 T} \cdot \sin 2\pi n f_0 t \Big|_{-\tau/2}^{\tau/2}$$

Now,

$$f_0 = 1/T$$

$$\Rightarrow \therefore a_n = \frac{2A\tau}{T} \cdot \frac{\sin(\pi n \tau/T)}{(\pi n \tau/T)}$$

As the function $x(t)$ is an *even* function, we can infer that there are no sine terms in the expansion and hence b_n must be 0.

The full Fourier series expansion can thus be written as:

$$\Rightarrow x(t) = A\tau/T + \sum_{n=1,3,5,\dots}^{\infty} \frac{2A\tau}{T} \cdot \frac{\sin(\pi n \tau/T)}{(\pi n \tau/T)} \cos(2\pi n f_0 t)$$

It is interesting to note that all of the Fourier series components for a train of pulses (data bits) with width τ (except for the dc component) are bounded by the

$$\frac{\sin(\pi n \tau/T)}{(\pi n \tau/T)} \quad \text{or} \quad \text{sinc}(n\tau/T) \quad \text{envelope.}$$

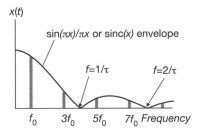

E EXAMPLE 1.3

Find the complex Fourier series expansion for the following waveform:

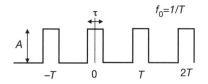

Solution

The general expression for the complex Fourier series expansion of a time waveform $x(t)$ is given by:

$$x(t) = \sum_{n=-\infty}^{\infty} c_n e^{j2\pi nf_0 t}$$

where

$$c_n = \frac{1}{T} \int_{-T/2}^{T/2} x(t) \cdot e^{-j2\pi nf_0 t} \cdot dt$$

For the pulse wavefrom shown above,

$$x(t) = \begin{cases} A \rightarrow -\tau/2 < t < \tau/2 \\ 0 \rightarrow \tau/2 < t < T - \tau/2 \end{cases}$$

Hence

$$c_n = \frac{1}{T} \int_{-\tau/2}^{\tau/2} A \cdot e^{-j2\pi nf_0 t} \cdot dt = \frac{A}{T} \cdot \frac{e^{-j2\pi nf_0 t}}{-j2\pi nf_0 t} \Big|_{-\tau/2}^{\tau/2}$$

$$\therefore c_n = \frac{A(e^{j2\pi nf_0(\tau/2)} - e^{-j2\pi nf_0(\tau/2)})}{j2\pi nf_0 T} = \frac{A\tau}{T} \cdot \frac{\sin(\pi nf_0\tau)}{\pi nf_0\tau}$$

Now $f_0 = 1/T$, hence

$$c_n = \frac{A\tau}{T} \cdot \frac{\sin(\pi n\tau/T)}{(\pi n\tau T)}$$

The full complex Fourier series expansion thus becomes:

$$x(t) = \sum_{n=-\infty}^{\infty} \frac{A\tau}{T} \cdot \frac{\sin(\pi n\tau/T)}{(\pi n\tau/T)} \cdot e^{j2\pi nf_0 t}$$

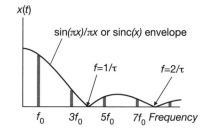

Spectrum of a data pulse

As the fundamental period of the time waveform increases, the fundamental frequency of the Fourier series components making up the waveform decreases and the harmonics become more closely spaced.

In the limit, as the time between pulses approaches infinity, the harmonic spacing becomes infinitely small and the spectrum is in fact continuous and bounded by the *sinc* function as shown.

A single pulse is not of course a periodic time function and the spectrum cannot strictly be evaluated using the Fourier series expansion. Instead the more general Fourier transform should be used.

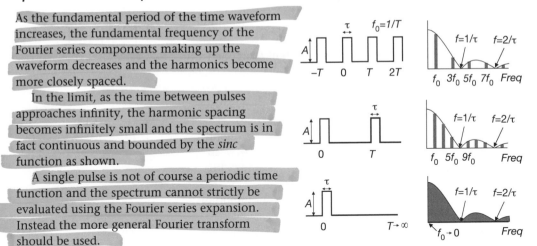

Spectrum of a baseband binary data stream

The spectrum (frequency domain representation) of a random data stream can be obtained by simply overlaying the instantaneous spectra for each individual pulse. We thus know that it will be bounded by the *sinc* envelope, and at any instance in time, the location and density of frequency components will depend on the particular pattern of data bits.

By shaping the data pulses, so that they have smooth edges, we would expect to reduce significantly the high frequency spectral content of the waveform. A commonly used pulse-shaping method is to pass the data stream through a low pass filter having a raised cosine response. The raised cosine filter belongs to a family of filters called Nyquist filters which have particularly useful properties in data communications (see Section 3.4).

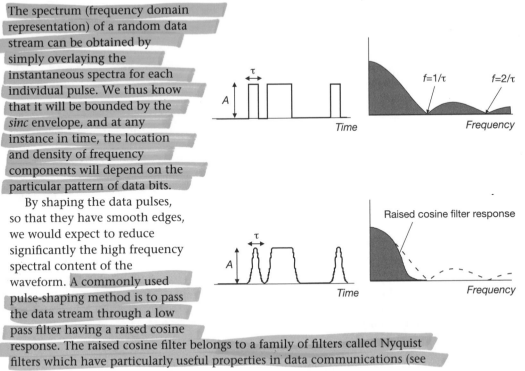

Factors affecting signal bandwidth

It is perhaps instinctively obvious that a waveform which has sharp transitions in the time domain will have a much higher harmonic content than one with smooth transitions. This is because the sharp changes in waveform can only be constructed from a large number of low-level high frequency sinusoids in a Fourier series expansion.

Hence, modulation formats that possess smooth pulse shapes or smooth phase transitions between symbol states are to be favoured when bandwidth is limited.

Not only is the shape of the waveform important in determining the amplitude of the frequency components within the Fourier series expansion, but the width of the data pulses also plays an important role.

As can be seen here, reducing the width of the pulse but keeping the period of the waveform constant results in an increase in the level of the higher harmonics at the expense of the lower harmonic levels. Overall, the energy content in the waveform has also gone down and so the combined power of the harmonics must also be reduced.

In the limit, as the pulse width tends to zero (that is, a delta function), we can expect the amplitude of each harmonic to approach a constant yet diminishing value.

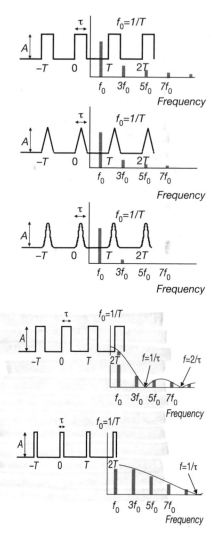

The Fourier transform

The *Fourier transform* is widely used for converting any mathematical description of a time domain waveform into the frequency domain equivalent. As such it can be viewed as a generalization of the Fourier series expansion. There is also an *inverse Fourier transform* which will convert from the frequency domain to the time domain.

The Fourier transform $X(f)$ of a time function $x(t)$ is defined as:

$$F\{x(t)\} = X(f) = \int_{t=-\infty}^{\infty} x(t)e^{-j2\pi ft} \cdot dt$$

and the inverse Fourier transform is given by:

$$x(t) = F^{-1}\{X(f)\} = \int_{-\infty}^{\infty} X(f)e^{j2\pi ft} \cdot df$$

A more detailed analyis of Fourier transforms can be found in Haykin (1989).

IN DEPTH

Fourier transform

Below is a table of common Fourier transform pairs that are often used in communications systems analysis.

Function	Time waveform	Spectrum
Constant	1	$\delta(f)$
Impulse	$\delta(t - t_0)$	$e^{-j2\pi f_0}$
Rectangular pulse	Width = T	$T\dfrac{\sin(\pi f T)}{\pi f T}$
Triangular pulse	Width = T	$T\left(\dfrac{\sin(\pi f T)}{\pi f T}\right)^2$
Cosine	$\cos(\omega_c + \phi)$	$0.5e^{j\phi}\delta(f - f_c) + 0.5e^{-j\phi}\delta(f + f_c)$
Impulse train	$\displaystyle\sum_{m=-\infty}^{\infty} \delta(t - mT)$	$f_0 \displaystyle\sum_{n=-\infty}^{\infty} \delta(f - nf_0)$ where $f_0 = 1/T$
Time delay	$x(t - T_d)$	$X(f)e^{-j2\pi f T_d}$
Conjugation	$x^*(t)$	$X^*(-f)$
Frequency translation	$x(t)e^{j2\pi f_c t}$	$V(f - f_c)$

1.2 Trigonometric relationships

The basic mixing process

In most designs of data modem, the frequency content of a baseband data stream (which we have just found using the Fourier series expansion) does not match the frequency transmission property of the transmission channel.

For example, a radio channel will have a *bandpass* response (see Section 4.5), only passing frequency components many times higher than those making up the input data stream.

In order to translate the spectrum of the input signal to fit within the passband of a channel, a process of modulation is employed as described in Chapter 5. This process often involves mixing the input data signal with a high frequency sine or cosine term, called the carrier.

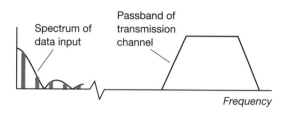

In order to understand how this mixing process achieves the desired goal, we need to revise the basic trigonometric relationships between sine and cosine terms.

The *carrier* term is usually represented mathematically as:

 $= \cos \omega_c t$

and we know that for the case of a 1,0,1,0,1,0,... data stream (see Example 1.1) the Fourier series expansion is:

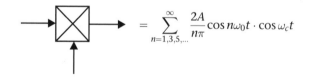 $= \sum_{n=1,3,5,...}^{\infty} \frac{2A}{n\pi} \cos n\omega_0 t$

The mixing process (assuming perfect hardware mixing components) simply performs a multiplication of the carrier term with each of the spectral components in the Fourier series expansion, thus the mixer output can be written as:

$$= \sum_{n=1,3,5,...}^{\infty} \frac{2A}{n\pi} \cos n\omega_0 t \cdot \cos \omega_c t$$

In order to evaluate this expression, we need to know the trigonometric expansion:

$$\cos A \cos B = \frac{1}{2}\cos(A - B) + \frac{1}{2}\cos(A + B)$$

Hence we can determine the output of the mixing process as:

$$Mixer\ output = \frac{1}{2} \cdot \frac{2A}{n\pi} \left(\sum_{n=1,3,5,...}^{\infty} \cos(\omega_c - n\omega_0)t + \sum_{n=1,3,5,...}^{\infty} \cos(\omega_c + n\omega_0)t \right)$$

If we plot the spectrum of the modulated signal, we see that it is centred on the carrier frequency, and for this example, reproduces the components of the baseband data signal exactly mirrored on either side of the carrier. We thus

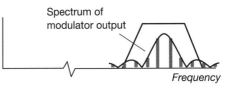

have a method to translate spectral components to any frequency we choose using the mixing process.

EXAMPLE 1.4

A square wave with a frequency of 1 MHz is mixed in a receiver with a local oscillator sinusoidal at 7.5 MHz and the resulting signal passed through a brick-wall low pass filter with a cut-off of 700 kHz.

(a) What will appear at the output of the receiver?

(b) The output of the receiver is found to be too small for practical use. How can this output level be increased simply by altering the shape of the 1 MHz modulating component?

Solution

The square wave is made up of sinusoidal components given by the Fourier series as derived in Example 1.1. This signal, when mixed with the 7.5 MHz local oscillator, will give components at the sum and difference between each of the Fourier series components and the 7.5 MHz reference.

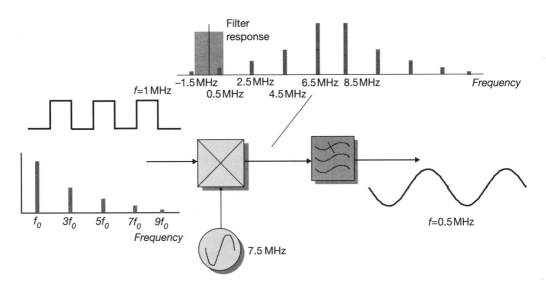

Only one of these components will fall within the bandwidth of the output low pass filter, hence the output waveform will be sinusoidal, with amplitude proportional to the amplitude of the seventh harmonic of the square wave.

In order to increase the output level from the filter, the amplitude of the seventh harmonic must be increased. This can be achieved by altering the mark space ratio of the square wave so that it becomes richer in harmonics (*see* Example 1.2).

Complex mixing processes

In more advanced modulation systems, or with more complex data patterns, combinations of sine and cosine components in the input signal may be mixed with combinations of sines and cosines in the carrier signals to give more control of the spectral content of the result. For this case, we need to know the full set of trigonometrical identities:

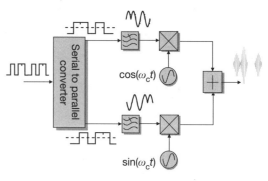

Complex modulation source

$$\cos A \cos B = \frac{1}{2}\cos(A - B) + \frac{1}{2}\cos(A + B)$$

$$\sin A \sin B = \frac{1}{2}\cos(A - B) - \frac{1}{2}\cos(A + B)$$

$$\sin A \cos B = \frac{1}{2}\sin(A - B) + \frac{1}{2}\sin(A + B)$$

$$\cos A \sin B = -\frac{1}{2}\sin(A - B) + \frac{1}{2}\sin(A + B)$$

The vector modulator

The arrangement of mixers and a combiner (summing device) shown here forms an extremely useful building block in digital communications systems. It achieves a linear frequency translation of all components in the input signal (represented by its in-phase and quadrature components) by a carrier frequency component (also represented by its in-phase and quadrature components). This building block is often referred to as a *vector modulator* or *quadrature modulator* and, as we shall see, can be used for both frequency *up-conversion* and *down-conversion*.

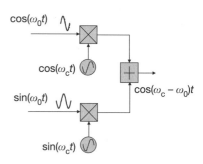

The output of the two mixing processes is given by:

$$\cos(\omega_0 t) \cdot \cos(\omega_c t) = 0.5 \cos(\omega_c + \omega_0)t + 0.5 \cos(\omega_c - \omega_0)t$$

and

$$\sin(\omega_0 t) \cdot \sin(\omega_c t) = -0.5 \cos(\omega_c + \omega_0)t + 0.5 \cos(\omega_c - \omega_0)t$$

which when subtracted from each other result in a single up-converted component:

$$\cos(\omega_c + \omega_0)t$$

and when summed give a down-converted component:

$$\cos(\omega_c - \omega_0)t$$

EXAMPLE 1.5

A vector modulator is fed with a perfect quadrature sinewave at the input, but there is a small phase error of 5° between the notional quadrature inputs of the carrier signal. What will be the ratio in dB between the sum and difference outputs of the vector modulator as a result of this phase error?

Solution

Let us write the inputs to the vector modulator as:

$$\cos(\omega_0)t, \quad \sin(\omega_0 t)$$

and the carrier inputs as:

$$\cos(\omega_c)t, \quad \sin(\omega_c t + \phi)$$

where ϕ is the phase error. Now:

$$\sin(\omega_c t + \phi) = \sin \omega_c t \cos \phi - \cos \omega_c t \sin \phi$$

and for small phase errors this can be approximated to:

$$\sin(\omega_c t + \phi) = \sin \omega_c t \cos \phi$$

The mixer outputs then become:

$$\cos(\omega_0 t) \cos(\omega_c t) = 0.5 \cos(\omega_c + \omega_0) + 0.5 \cos(\omega_c - \omega_0)t$$

$$\sin(\omega_0 t) \sin(\omega_c t + \phi) = -0.5 \cos(\omega_c + \omega_0)t \cos \phi + 0.5 \cos(\omega_c - \omega_0)t \cos \phi$$

At the output of the summing device we get a wanted term at the difference frequency and an unwanted term (usually referred to as the image) at the sum frequency as follows:

Difference term:

$$0.5[1 + \cos\phi]\cos(\omega_c - \omega_0)t$$

Sum term:

$$0.5[1 - \cos\phi]\cos(\omega_c + \omega_0)t$$

The ratio of the amplitude of the wanted to unwanted terms is thus:

$$\textit{Amplitude ratio (image suppression)} = \frac{[1 + \cos\phi]}{[1 - \cos\phi]}$$

For a phase error of 5°, the amplitude ratio of wanted to unwanted signals is thus 525:1, or a relative power level of approximately 27 dB.

IN DEPTH

Useful trigonometrical identities

The table below gives a list of useful trigonometrical identities which are often exploited in the design of digital communications circuits.

$$e^{\pm j\theta} = \cos\theta \pm j\cdot\sin\theta$$
$$\cos\theta = (e^{j\theta} + e^{-j\theta})/2$$
$$\sin\theta = (e^{j\theta} - e^{-j\theta})/2j$$
$$\sin^2\theta + \cos^2\theta = 1$$
$$\cos 2\theta = \cos^2\theta - \sin^2\theta$$
$$\sin 2\theta = 2\sin\theta\cos\theta$$

1.3 Communications networks and signalling protocols

What is a network?

While this book is not intended to be a definitive text on the networking aspects of communications systems, this section has been included to allow the main focus of the book – the design and performance of the modem – to be appreciated within the wider context of its place in the communications network. A very good text on networking has been written by Halsall (1992).

The *network* is the all-embracing term for the collection of building blocks that make up a modern sophisticated communications system. It in general comprises *physical interconnections* via cable, fibre, radio or infrared, *modems* which process the information for reliable transmission through a given type of interconnection, and *switches* (routers, exchanges) which are used to route the information between source and destination.

The end-user equipment, such as the telephone, fax or computer, is not usually considered to be part of the network, but rather a terminal which 'plugs into' the network.

A typical network configuration

Shown here is a typical configuration for a modern cellular telephone network. The wireless connection can be seen to be only a small part of a much larger network involving: a *mobile switching centre* to route calls from mobile to mobile or into the exchange; provision of private cable or *microwave radio links* for the interconnection of *base-station* sites to switching centres and between switching centres; interconnection to the *public switched telephone network (PSTN)* for the transfer of calls between mobile and domestic telephones.

Network hierarchy

The PSTN provides a good example of transmission hierarchy within a network, where the connections carrying little traffic are served with low capacity links (for example, the twisted pair cable connection to the average home which can support up to 56 kbps with a modem, or 2 × 64 kbps with a special ISDN (Integrated Services Digital Network) line), through to the interconnections between the major cities which often comprise very high speed optical fibre links carrying in excess of 500 Mbps and supporting over 7000 voice calls each.

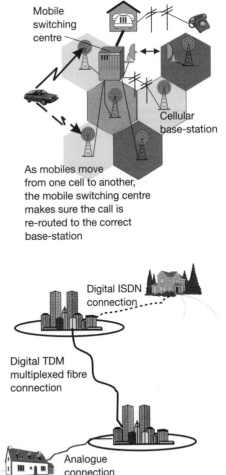

Transmission standards

In most countries in the world, a set of industry standards exist for the transmission rate and format over different parts of the network to ensure compatibility of signalling and switching equipment at each end.

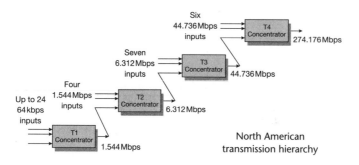

North American transmission hierarchy

North America, Europe and Japan all have slightly different signalling rates for parts of their network hierarchy, with the North American standard illustrated here (see the in-depth section for the equivalent European standard).

IN DEPTH

Frame structure and multiplexing hierarchy for European ITU (CCITT) telephony

In Chapter 7, the method by which individual voice channels are digitized at the local telephone exchange or switching centre is discussed. Each incoming analogue phone line is sampled at a rate of 8000 samples per second, and each sample is represented as an 8-bit word.

In order to send these digitized voice samples between switching centres, high capacity data links are used, often optical fibres, capable of supporting several gigabits/second. The individual 8-bit words are grouped into frames for transmission as shown below, with the addition of an 8-bit signalling word and an 8-bit framing/synchronization word. Thirty-two 8-bit words are used in the E-series ITU standard, and twenty-four 8-bit words in the American T-series standard.

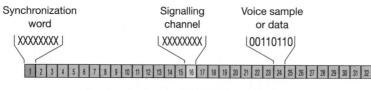

Framing structure for ITU (CCITT) multiplexing

An E-series frame is thus $8 \times 32 = 256$ bits wide, and is sent at a rate of 8000 frames per second, to match the voice channel sampling rate. The bit rate on the basic E1 channel is thus $256 \times 8000 = 2048$ Mbps.

E1 channels can be grouped into super frames to give E2 channels, E2 into E3 channels and so on. Additional framing and signalling words are added at each stage to control routing within the switching network.

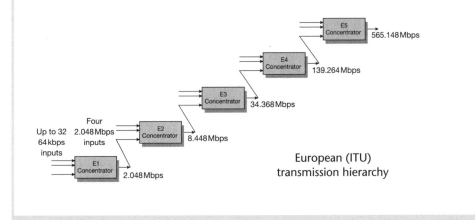

European (ITU)
transmission hierarchy

Integrated Services Digital Network (ISDN)

The ISDN network is a fully digital communications link standard for interconnection between the local exchange and the home or office. It provides a guaranteed data rate of a minimum of 64 kbps and does away with the need for separate modem cards within computers or fax machines. Voice is digitized at the source (that is, within the telephone) and so there is perfect reception quality under all conditions. It is likely that ISDN will replace many of the existing analogue connections to the user over the next few years.

The ISDN system also has a separate set of service standards with the '2B+D' service intended for small office and home use, and the 'primary rate' services intended for major business use.

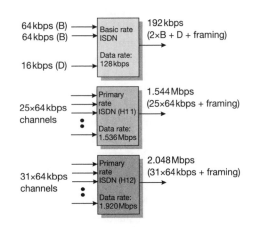

Standard ISDN services

What is a protocol?

In order for information to be correctly and efficiently routed from source to destination within a network, it is necessary for the path through the network (that is, the switch settings) to be set appropriately.

The signalling between source, switching centres and destination used to set up the route must be understood by all parties if it is to operate successfully. In other words, it must conform to a pre-established *protocol*. This protocol usually takes the form at the lowest level of sending information (data words, or perhaps tones (DTMF)) alongside the message signal which could convey information about the destination address, the type of message, the message length, the occurrence of transmission errors, and so on, to be picked up by the switching centres or terminal equipment as appropriate.

There are many different protocols used within large communications networks, each hopefully optimized for a given control task. For example, the protocol used to control information flow over a wireless portion of a network often needs to be more robust at detecting and correcting errors than a protocol employed on a good quality wired link. A well-known protocol for computer interconnection is the RS232 protocol. A commonly used packet switch protocol is the X.25 standard.

IN DEPTH

The RS232 protocol

The RS232 standard was drawn up by the CCITT in Europe and the Electronics Industries Association (EIA) in the United States in order to ensure that there was a worldwide common format for serial communication between computers and peripherals. This standard outlines the characteristics of the connectors to be used, (size, number of pins, shape and so on), the voltage levels to be supported, and the control functions assigned to each pin. It also identifies the simple handshaking protocol which determines when the terminal equipment is ready to send and to receive data.

The voltage levels for the RS232 interface are defined to lie between $-3\,V$ and $-15\,V$ for logic 1, and $+3\,V$ and $+15\,V$ for logic 0. Typically they are set to $-12\,V$ and $+12\,V$. Note that the RS232 interface uses so-called 'negative logic'.

The interface is normally limited to speeds of 20 kbps over distances in the order of 15 metres due to pulse rise times over these lengths of cable. Higher speeds are possible over much shorter distances, however.

A similar standard to the RS232, termed RS449, can deliver much higher interconnection speeds by specifying the use of 'balanced signalling' and much tighter specification of interconnection cable parameters. The RS449 interface can signal at speeds in excess of 2 Mbps.

A table of pin designations for the RS232 interface is given below.

1	Ground	14	Secondary transmitted data
2	Transmitted data (TD)	15	Secondary transmitted clock
3	Received data (RD)	16	Secondary received data
4	Request to send (RTS)	17	Receiver clock (RC)
5	Clear to send (CTS)	18	Divided clock receiver (DCR)
6	Data set ready (DSR)	19	Secondary request to send
7	Ground	20	Data terminal ready (DTR)
8	Data carrier detect (DCD)	21	Signal quality detect (SQ)
9	Not connected	22	Ring indicator (RI)
10	Not connected	23	Data rate selector
11	Not connected	24	External transmitter clock
12	Secondary data carrier detect	25	Busy
13	Secondary clear to send		

The seven-layer OSI model

It is often the case that there are several types of signalling protocol running concurrently to control the information flow and processing within a network – some dealing with information on a bit-by-bit basis, such as error correction protocols, and some operating at a much higher level to ensure that whole files of data are packaged correctly and arrive in sequence after transmission.

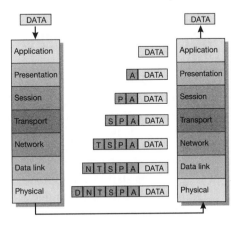

OSI network model

These various levels of protocol are often categorized in terms of their function within a now standardized seven-layer model, known as the *OSI seven-layer model*, as shown here.

As the data passes through each layer of the network hierarchy, additional signalling information is usually added to pass information to the corresponding layers at the receive side. This can amount to a very large overhead for the system, particularly if the data packet size itself is small.

Network types: circuit and packet switched network operation

Most modern networks operate in either a circuit or packet switched mode.

A circuit switched network is one where at the start of each complete message transaction (for example, a telephone call), the route through the network is identified, the correct links are switched in, and this configuration is held for the entire duration of the call.

A packet switched network, on the other hand, routes small chunks (*packets*) of the message, hopefully down the best route (least congested, least noisy and so on) available at the time.

Packet switched networks are able to 'optimize' the routing of data on a packet-by-packet basis and hence these systems are very attractive for ensuring the best use of the network capacity – particularly for unpredictable densities and duration of traffic such as Internet access. The protocol overhead for routing of individual packets, and for ensuring that packets arrive on time and in the right sequence when they have travelled by several different paths, makes the circuit switched approach more attractive for some applications such as voice traffic. There is, however, a trend towards using packet switched networks for voice also.

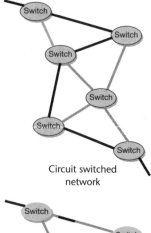

Circuit switched network

Packet switched network

Network management

A major factor in any communications network is the management of the network, which covers not only the intelligent routing of traffic around the network, making the most efficient use of the resource, but also taking care of generating billing information for charging users, and gathering statistical information for monitoring the network performance and loading.

Today, the major task of network management is automated and realized through millions of lines of computer code running on powerful computers within the main switching centres.

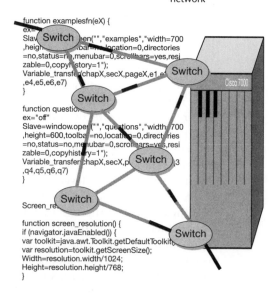

The network management task of efficient routing within a mobile network is particularly challenging, with the network having to know and track the location of several hundreds of millions of mobile users who can be anywhere in the world and then select the most cost-effective, or least utilized, or minimum time delay route between the calling parties!

Network link layer

The *data link layer* as defined in the seven-layer OSI model is focused on achieving reliable information transfer over the *physical layer* (the data modulation/demodulation process). This link layer includes the functions of error detection and message/packet retransmission.

Two types of link layer operation are usually defined – *connectionless* and *connection orientated*.

Connectionless operation launches frames or packets into the network to be transferred to the destination, but with no tracking of the frames to ensure correct delivery; that is, if an error is detected in the frame or packet and it is discarded, there is no means of provision for a retransmission to occur.

Connection-orientated operation, on the other hand, has all the built-in tracking mechanisms and facilities for retransmission to make sure that guaranteed error-free transmission occurs.

Most of the remainder of this book is concerned with the physical layer of the OSI model – the means by which binary data is best communicated over the physical channel. Chapter 7, however, comes back to discuss some of the error detection and retransmission issues involved in the link layer.

Synchronous/asynchronous communications

Synchronous transmission

A *synchronous* system is one in which the transmitter and receiver are operating continuously at the same number of symbols per second and are maintained, by suitable correction, in the desired phase relationship.

Synchronous operation requires an accurate timing signal in the receive modem, which can be derived either from a separate transmitted reference or from symbol transitions within the data signal itself. A separate reference requires additional power or bandwidth, while a data-derived reference requires frequent symbol transitions to occur in the received waveform; that is, long strings of 1s or 0s are not acceptable.

Asynchronous transmission

An *asynchronous* system is one in which the symbol rate can vary marginally with time and no rigid timing constraint is applied.

Asynchronous operation is usually characterized by the use of 'start' and 'stop' bits to signify the beginning and end of a character that is to be sent, for example the RS232 protocol.

ASCII character format

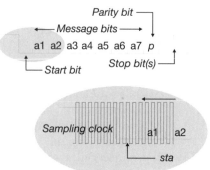

The asynchronous method of transmission is very effective when the information to be sent is generated at irregular intervals, for example entering letters from a keyboard, where there is no prior knowledge of when the keys will be pressed. The data receiver must therefore be alerted to the fact that a new data message is arriving and also when the message ends so that it can enter an idle mode (often called marking).

Shown here is the common format used for sending ASCII characters over a modem link. When the link is inactive, the data line is kept high; it is pulled low for one bit period to bring the data receiver out of its idle mode. This is called the 'start bit'. Following the start bit are the seven bits representing the ASCII character set, followed by a parity check bit. Finally, one or more stop bits are added forcing the data line high, mimicking the idle state, such that the receiver is then ready to detect the presence of a new start bit.

In an asynchronous receiver, the data bits are sampled at a rate many times the bit rate with typically 16 clock cycles per data bit. Once a transition from high to low representing a start bit is detected, the data is then sampled on the eighth clock cycle, with the expectation that this will be close to the centre of the bit period and hence a reliable measure.

✔ ### Advantages of synchronous data communications

- Superior noise immunity due to *matched filtering*; that is, the symbol or bit is averaged over its entire duration giving optimum noise and interference rejection and maximizing signal power.

- Can accommodate higher data rates than asynchronous systems.

✘ ### Disadvantages of synchronous data communications

- Requires finite time for synchronization to occur.

- Is more expensive and complex than asynchronous operation.

- Cannot easily accommodate variable symbol rates.

1.4 Definition of terms

This section introduces many of the terms and definitions relating to the communications channel. Some of these terms can have specific meanings within the context of a digital communications system, and these are defined here.

Characteristics of message type

Analogue

An analogue signal is defined as a physical time-varying quantity and is usually smooth and continuous, for example acoustic pressure variation when speaking. The performance of an analogue communications system is often specified in terms of its *fidelity* or *quality*, hence the term 'hifi' – HIgh FIdelity.

Digital

A digital signal, on the other hand, is made up of discrete symbols selected from a finite set, for example letters from the alphabet or binary data.

The performance of a digital system is specified in terms of accuracy of transmission, for example *bit error rate (BER)* and *symbol error rate (SER)*.

Elements of a communications link

Transmitter

The transmitter (TX) element processes a message signal in order to produce a signal most likely to pass reliably and efficiently through the channel. This usually involves modulation of a carrier signal by the message signal (see Chapters 5 and 6), coding of the signal to help correct for transmission errors (see Chapter 7), filtering of the message or modulated signal to constrain the occupied bandwidth (see Chapter 3), and power amplification to overcome channel losses.

Transmission channel

Loosely defined as the electrical medium between source and destination, for example cable, optical fibre or free space, the channel is characterized by its loss/attenuation, bandwidth, noise/interference and distortion.

Receiver

The receiver (RX) function is principally to *reverse* the modulation processing of the transmitter in order to recover the message signal, attempting to compensate for any signal degradation introduced by the channel. This will normally involve amplification, filtering, demodulation and decoding and in general is a more complex task than the transmit processing.

Sources of link degradation

Distortion

The common types of link distortion are:

- frequency-dependent phase shifts, giving rise to differential group delay (see Section 4.2)
- gain variations with frequency caused by the channel filtering effect (see Section 4.1)
- gain variations with time as experienced in a radio/infrared channel
- frequency offsets between transmitter and receiver due to Doppler shift or local oscillator errors (see Section 4.3).

   Distortion can be introduced within the transmitter, the receiver and the channel. In some cases it can be corrected using channel equalizers (see Section 4.4), and gain and frequency control systems (see Section 4.2). Unlike noise and interference, distortion disappears when the signal is turned off.

Interference

Interference arises owing to contamination of the channel by extraneous signals, for example from power lines, machinery, ignition systems, other channel users and so on. If the characteristics are known, then interference can often be suppressed by filtering or subtraction (for example, car suppressers).

   Interference is often impulse-like in nature and we know from our knowledge of the Fourier transform and Fourier series expansion (Section 1.1) that an impulse contains energy over a very wide bandwidth. In the case of ignition noise, the ignition system may be firing at only 4000 Hz (1000 rpm), yet significant high frequency energy will exist at frequencies of several MHz.

Noise

Noise is charcterized as *random, unpredictable electrical signals from natural sources,* for example atmospheric noise, thermal noise, shot noise and so on.

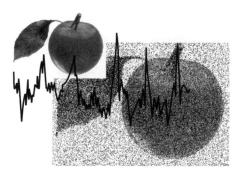

Because of the multiplicity of noise sources in a communications link, it is difficult to define the properties (frequency range, level and instantaneous phase) of noise and hence equally difficult to reduce its effect on the communications link performance. For convenience, most textbooks and indeed practising engineers assume noise to fall predominantly into the class of Additive White Gaussian Noise (AWGN) which does indeed form an adequate classification for most cases. However, we should not forget that this is a *general simplification* of the whole noise issue.

Transmission protocols

Simplex

A simplex communications link is defined as one where the communication flow can only occur in one direction. A common example is broadcast radio or television.

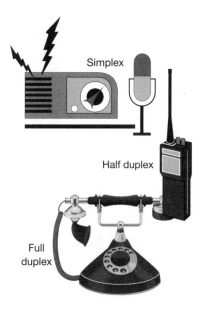

Half duplex

A half-duplex communication link is capable of communication in both directions, but not at the same time. This form of communication is commonly employed in applications from 'walkie-talkies' and CB-radio through to professional police radio systems.

Full duplex

A full-duplex system, as one would expect, can support simultaneous two-way communication. The most widespread example is the telephone.

Unipolar vs bipolar waveforms

When signalling over a communications link, there are two common binary (two state) signalling formats, *unipolar* and *bipolar*.

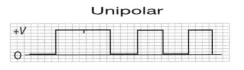

Unipolar

A unipolar scheme is characterized by the voltage states being zero and $+V$ volts, and thus has a dc component to the Fourier series expansion (Section 1.1).

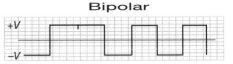

Bipolar

A bipolar format has a zero dc mean with voltage states of $+V$ and $-V$ volts representing logic 1 and logic 0. The bipolar format is used on the serial output port of every computer conforming to the RS232 interface standard (see Section 1.3) having voltage states of $+12\,V$ and $-12\,V$.

In the section on matched filtering in Chapter 3, it is shown that this bipolar format of signalling for data transmission is much more tolerant to noise than the unipolar equivalent for the *same average symbol power*.

Q QUESTIONS

1.1 Find the trigonometrical Fourier series expansion for the following waveform:

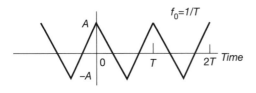

1.2 Find the trigonometrical Fourier series expansion for the following waveform:

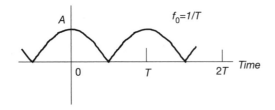

1.3 Find the trigonometrical Fourier series expansion for the following waveform:

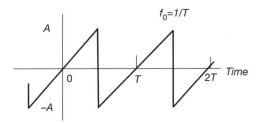

1.4 The data waveform shown here can be seen to contain a long string of 1s as part of the data pattern. What effect will this have on the frequency content of the signal during this period?

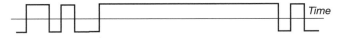

1.5 A square wave has a maximum and minimum voltage of $+2.5\,V$ and $-2.5\,V$ respectively. What is the dc value and the level of the first three Fourier series components?

1.6 A 1200 bps binary data signal with an alternate $1,0,1,0,1,\ldots$ bit pattern is input to a spectrum analyser. At what frequencies will components appear on the analyser trace?

1.7 A communications link is sending data at 9600 bps by using four different voltage levels, each representing a pair of information bits, so that the voltage levels on the line only change at half the required data rate. What is the minimum bandwidth required to pass the fundamental frequency of this four-level waveform?

1.8 A mixer has two inputs, one a square wave data signal with fundamental frequency of 600 Hz, and the other a carrier wave with frequency of 10 MHz. The carrier wave, however, is not a perfect sinusoid, but is partly limiting at the top and bottom of the waveform such that it contains a second harmonic component at a level of one-tenth of the fundamental.

Sketch the output from the mixing process, assuming that the square wave input is limited to the first five non-zero harmonics only.

1.9 An otherwise ideal mixer generates an internal dc offset which sums with the data input signal. How will this affect the spectrum of the output signal for a perfect carrier sinewave input?

1.10 A vector modulator is fed with a perfect quadrature sinewave at the input, but there is a small amplitude imbalance of 0.1 dB between the voltage levels of the otherwise perfect quadrature carrier signals. What will be the ratio in dB between the sum and difference outputs of the vector modulator as a result of this amplitude error?

2 Data transmission fundamentals

Chapter 2 is all about identifying the factors which are going to dominate in the early stages of a digital communications system design. Without a good understanding of the 'big issues' of bandwidth and noise in dictating data throughput and modem complexity, the more detailed issues of modulation type, hardware implementation and performance optimization, covered in the later chapters, will not fit into place.

Multi-output comparator

Most of the discussion centres on the engineering aspects and how they influence system design; however, the first section covers issues which *fall outside the control of the engineer*, such as regulatory or commercial pressures, yet which often have a significant part to play in dictating the final engineering solution.

Section 2.2 begins to unravel how *channel bandwidth* and *system noise* dominate the performance of all digital communication systems, demonstrating that the designer is always faced with a trade-off between the two. Section 2.3 takes us quickly away from binary signals of 1s and 0s to multi-level modulation, where the signal on the channel can represent a single bit or thousands of bits at any one time, as the designer chooses. Finally, Section 2.4 ties up the loose ends of the by-now-familiar noise and bandwidth debate, introducing the theoretical optimum performance for a digital communications link as defined in the *Shannon bound*.

2.1 Factors affecting system design

Technological limitations

Hardware and software availability

It is often the case in engineering that we know of an optimum design of a system, but do not yet have the technology invented or sufficient processing power to implement it.

A recent example in data communications is the new digital cellular system GSM, where the modulation format GMSK (Gaussian Minimum Shift Keying) was chosen, rather than the technically superior QPSK format (Section 6.4), owing to the problems with realizing a cost- and power-efficient linear power amplifier required by QPSK in the handsets. (Subsequently, suitable amplifiers have been invented and the latest cellular system proposals use QPSK.)

Digital signal processors (very fast special-purpose microprocessors) are now being used in many applications to implement functions traditionally realized using hardware components. In fact most high-speed telephone modems are implemented in digital signal processing. The processing power of these devices is increasing approximately 2–3 fold every year.

IN DEPTH

Digital signal processing

In 1979, Intel introduced the first microprocessor with an architecture and instruction set specifically tailored to digital signal processing (DSP) applications. Since then, general purpose DSP chips have been launched by Texas Instruments, IBM, Analog Devices, Motorola, Inmos and Lucent (AT&T) among others, and DSP ASIC (application-specific integrated circuits) cores are available from these manufacturers and many others, in particular the OAK DSP core.

The rapid growth in the exploitation of DSP in digital communications is not surprising considering the commercial advantages now offered by their low cost and ease of programmability.

Modern DSP devices, for example the TMS 320C6X series, are extremely powerful, able to implement the modem, error correction, channel equalization and voice coding functions required in a modern digital cellular phone within a single device, and potentially several times over. Some of the basic benchmarks for this processor are shown in the table below.

TMS 320C6201 DSP	
Algorithm types	**Execution time**
FFT (256 points)	13.3 μs
Viterbi decoder for GSM ($N = 189$)	36.2 μs
Linear phase filter	0.29 μs
Infinite impulse response filter (8 biquads)	0.24 μs

As an example, this processor can implement a raised cosine filter with 50 coefficients (taps), which would typically be used in a M-ary QAM data modem, within approximately 0.15 μs and could thus accommodate a data symbol rate of about 3 000 000 symbols per second.

Power consumption

The power consumption/performance trade-off is almost always a consideration for engineers, especially if the equipment is to be battery powered. Using the same example of cellular handset power amplifiers, the problem is not only to achieve linearity in the amplifier, but to do so with sensible power efficiency. At present, approximately half of the power drawn from the battery in a handset is wasted in heat in the RF amplifier.

Component size

Sizes of electronic components are of course diminishing, but so is the space into which engineers are expected to squeeze very complex circuitry. Complete wireless data modems are beginning to appear conforming to the now familiar PCMCIA 'credit card' format for computer peripherals. Here, the problem is not only how to fit radio frequency and digital processing into such a small space, but also how to accommodate a decent aerial!

Government regulations and standards

In communications, perhaps more than any other field, the need for standards to ensure correct interoperation of equipment is paramount. Most items of equipment that we use in everyday life, such as kettles, washing machines and so on, operate independently of other items of equipment and so standards are not too important. Communications equipment, on the other hand, is always interworking with other devices, possibly located on the other side of the world.

The drawing up of standards falls to a small number of national and international bodies, with, for example, ETSI (European Telecommunications Standards Institute) being responsible for the drafting of most of the new wireless communications standards for Europe, and ITU (International Telecommunications Union) providing the same function for wired communications equipment such as telephone/computer modems. Policing of these standards falls usually to national agencies. For example, all equipment to be connected to the UK telephone network must be BABT (British Approvals Board for Telecommunications) approved to ensure compliance with the standards.

With wireless communications, not only is it necessary to ensure interoperability of equipment, it is also necessary to specify radiation parameters – power level, occupied bandwidth and so on – in order to ensure that interference in not caused to other users. Where possible, radio frequency allocations are agreed on a global scale at the World Administrative Radio Conference (WARC) held every five years.

Commercial realities

The reality of the communications market-place is that cost and appearance mean more to the consumer than the technology inside. Mobile phones sell on their style and talk time, rather than their receiver sensitivity or BER performance. This is an important lesson for engineers to grasp, for although some mobile phones are much better technically than others, all are simply assumed by the customer to work properly, and achieving technical excellence may not lead to excellent sales.

It is a difficult and challenging task to design a product that meets user expectations and needs without *over-engineering or under-engineering*, and all the cost, timescale and reliability issues that follow. It is hoped that this book will provide a good grasp of the design choices available to the digital communications engineer, from which a correctly engineered product can be achieved.

2.2 Data transmission fundamentals

How quickly can information flow?

Having covered the 'non-engineering' factors affecting data transmission in Section 2.1, we now focus on establishing the ground rules governing effective engineering design of a digital communications link. To do this, it is helpful to consider the question: 'What limits how quickly information can be sent over any given channel?'

Given a requirement to send digital information from a source (for example, a computer) to a destination (for example, a printer), let us consider how quickly we can transfer 1 000 000 bits of information over a communications link:

- More than 1 second?
- More than 1 millisecond?
- Instantaneously?

By moving on to look at some possible transmission methods we will identify the critical limiting factors and hence determine the answer.

Methods of communication

Binary signalling

- **Binary signalling using a single cable**
  Here the transmission rate is determined by how
  fast the voltage (or other symbol type) can be
  varied on the channel, before the frequency
  content (as predicted by the Fourier series
  expansion (see Section 1.1)) is so high that the
  inevitable filtering of the channel attenuates and hence distorts too much
  of the signal. In other words, it is limited by the *bandwidth* of the link.

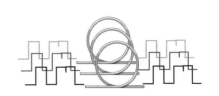

- **Binary signalling using many parallel cables**
  By using multiple cables, the throughput over the
  link can be increased in proportion to the number
  of cables (channels) used. Alternatively, the
  throughput can be maintained at that of the single
  binary link, allowing lower bandwidth (probably
  lower cost) links to be substituted.

Multi-level signalling

- **Multi-level signalling using a single cable**
  There is no reason why data transmission
  should be limited to a binary (two symbol
  state) format over a channel, and in theory,
  it should be possible to use any number of
  voltage levels or symbol types.

    For example, using four voltage levels means that we can uniquely
  encode two bits into each of the four levels (00 = level A, 01 = level B,
  10 = level C, 11 = level D). This means that every time we change the
  symbol state, two bits of information are conveyed compared with only
  one for the binary system. Hence, we can send information *twice as fast* for
  a given bandwidth of link, or use a link with *half the bandwidth* and
  maintain equivalent transmission rate.

- **Multi-level signalling using multiple cables**
  It is of course possible to use multi-level
  signalling (often termed *M-ary signalling*)
  over parallel channels if so desired, with the
  consequent increase in throughput or
  opportunity to reduce the bandwidth on
  each channel as required.

Multi-level symbol operation

In principle we can use any number of symbols (symbol states) for conveying digital information. For example, why not use 1024 different voltage states, each state (symbol) conveying $\log_2 1024 = 10$ bits? We could even consider using 1 048 576 symbol states, with each symbol conveying 20 bits of information!

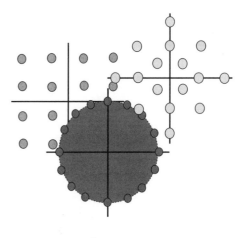

Clearly there is a practical limit on the number of states to be used, governed by the ability of the receiving equipment to *accurately resolve* the individual states (voltage levels, frequencies, light intensities and so on). This will be determined principally by the levels of noise and distortion introduced by the channel and by the TX and RX units.

For example, some of the more recent telephone modems operating at 56 kbps use in excess of 1024 different symbol states (combinations of amplitude and phase of carrier) to signal over the telephone channel, while the current digital cellular telephone systems use only two or four states because the equipment has to operate in much noisier (electrically) environments.

The bandwidth and noise trade-off

Having briefly considered the question of 'how fast can we send data', it is immediately apparent that there are two fundamental factors affecting the information transfer rate on a channel, namely:

- The maximum possible detectable rate of change of waveform or symbol state
  ▸ The *bandwidth* of the channel (and any bandwidth limits imposed by the transmitting and receiving devices) will determine how quickly the signalling states on the channel can be changed.

- The ability to resolve any number of discrete symbol states
  ▸ The level of *noise* in the channel will impose an upper limit on the number of different unique symbol states that can be correctly resolved (decoded) at a receiver.
  ▸ The degree of *distortion* introduced by the channel will also limit the number and rate of change of symbol states that can be accommodated with acceptable performance.

So, if we had a channel with infinite bandwidth, or no noise and distortion, it would be possible to send the 1 000 000 bits of information 'instantaneously' – well, at the speed of light anyway.

Information transfer rate

> The *information transfer rate* for a data channel is defined as the speed at which *binary information (bits)* can be transferred from source to destination.
>
> Units of *Information transfer rate* → *bits/second*

For example, if six bits of information are sent every 6 ms, then,

Information transfer rate = 6 bits/6 ms = 1000 bits/second

Symbol rate (baud rate)

The information transfer rate must not be confused with the rate at which symbols are varied to convey the binary information over the channel. We already know that we can encode several bits in each symbol.

> The correct definition of *symbol rate* (sometimes called *baud rate*) is the rate at which the signal state changes when observed in the communications channel and is not necessarily equal to the information transfer rate.
>
> Units of *Symbol rate* → *symbols/second* (*baud*)

For example, if a system uses four frequencies to convey pairs of bits through a channel, and the frequency (symbol) is changed every 0.5 ms, then:

Symbol rate = 1/0.5 ms = 2000 symbols/second (2000 baud)

The information transfer rate for this example, however, is 4000 bps, as each symbol conveys two bits.

Bandwidth efficiency

> The *bandwidth efficiency* of a communications link is a measure of how well a particular modulation format (and coding scheme) is making use of the available bandwidth.
>
> Units for bandwidth efficiency of a digital communications link
>
> *Bandwidth efficiency → bits/second/Hz*

For example, if a system requires 4 kHz of bandwidth to continuously send 8000 bps of information, the bandwidth efficiency of the link is:

Bandwidth efficiency = 8000 bps/4000 Hz = 2 bits/second/Hz

2.3 Multi-level signalling (M-ary signalling)

The relationship between *bits* and *symbols*

It is now very uncommon to design modems that use only binary (two symbol) signalling, with users demanding ever higher data rates in the same channel bandwidth. It has already been mentioned that some modern dial-up modems use over 1024 signalling states.

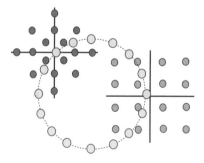

> The number of symbol states needed to uniquely represent any pattern of *n* bits is given by the simple expression:
>
> $M = 2^n$ *symbol states*

For example, a group of three bits can be represented by one of:

$M = 2^3 = 8$ symbol states
4 bits by $M = 2^4 = 16$ symbol states
5 bits by $M = 2^5 = 32$ symbol states

and so on.

Example: 8-ary signalling

The purpose of using multi-level or multi-symbol signalling is to allow the designer to trade channel capacity with bandwidth and noise immunity. Consider, for example, a system employing eight voltage states rather than a simple binary two-state design.

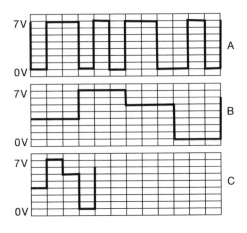

Trace A represents the binary data source to be encoded into the 8-ary signal.

Trace B is the encoded signal with the *information rate* kept the same for both binary and 8-ary systems. The result is that the rate at which the voltage state is varied on the channel is reduced by a factor of three. This translates directly into a threefold reduction in bandwidth required to support communication.

Finally, trace C shows an 8-ary signal which has the same *symbol rate* as the binary source and hence requires the same bandwidth, but the information rate has been increased threefold.

E EXAMPLE 2.1

A modem claims to operate with a bandwidth efficiency of 5 bits/second/Hz when using 1024 symbol states in the transmission constellation.

(a) How many bits are being encoded in each symbol, and what is the modem capacity if the baud rate is 4000 symbols/second?

(b) How many symbol states must be employed if the user wishes to send his information in half the time?

Solution

(a) For 1024 symbol states, the number of bits represented by each symbol is $\log_2 1024 = 10$ bits/symbol.
    For a baud rate of 4000, this means that the information transfer rate is $4000 \times 10 = 40$ kbps.

(b) In order to send the information in half the time, it would be necessary to send data at 80 kbps and hence to encode 20 bits in each symbol. The number of symbol states is thus a massive $2^{20} = 1\,048\,576$.

✔ **Advantages of M-ary signalling – summary**

- A higher information transfer rate is possible for a given symbol rate and corresponding channel bandwidth,

or

- A lower symbol rate can be obtained, leading to a reduced bandwidth requirement for a given information transfer rate. (Both result in an increase of *bandwidth efficiency* – bits/second/Hz.)

✘ **Disadvantages of M-ary signalling – summary**

- M-ary baseband signalling results in reduced noise/interference immunity when compared with binary signalling (see Section 3.6), as it becomes more and more difficult to distinguish between symbol states.

- It involves more complex symbol recovery processing in the receiver (see Section 3.5).

- It imposes a greater requirement for linearity (see Section 4.3) and/or reduced distortion in the TX/RX hardware and in the channel (except for orthogonal M-FSK (see Section 6.3)).

2.4 Calculation of channel capacity

Limitation due to finite bandwidth

In order to determine the maximum rate at which data can be sent over a channel, we need to know the maximum symbol rate that can be supported in a channel as a function of the channel bandwidth.

For the moment, let us consider only 'low pass' or 'baseband' channels where we can assume that the channel can pass signals with frequencies within the range 0 Hz to B Hz. This is called the *channel passband*. Shown here is an example of an 8-ary symbol stream which just happens to begin to alternate between the maximum and minimum voltage levels. This looks like a square wave, for which the harmonic structure is known from its Fourier series expansion (see Section 1.1). The fundamental of this square wave is at a frequency of $0.5 \times 1/T_s$, where T_s is the symbol period.

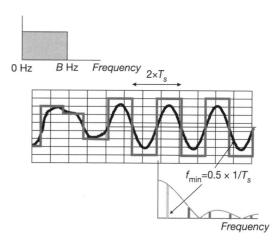

Consider what happens if the channel has only sufficient bandwidth to pass the fundamental of the square wave.

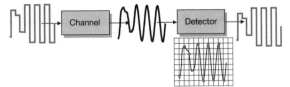

Providing that we are careful to maintain accurate levels throughout the system, and we sample the received signal at the correct time within each symbol period, then it is possible to recover the symbols and hence the data with a channel that has a passband of only $0.5 \times 1/T_s$ Hz.

Because the square wave signal chosen represents the maximum and most extreme rate and range of signal change for the 8-ary example, we can simply infer that any other symbol pattern will require less bandwidth for transmission and hence $0.5 \times 1/T_s$ is indeed sufficient bandwidth for all cases. It is also the *minimum bandwidth* needed as any less would result in the fundamental of the square wave being suppressed and no signal getting through the channel.

Minimum transmission bandwidth

From the simple 8-ary example, we can generalize and conclude that:

> The minimum bandwidth required for error-free transmission in a *baseband* channel is given by:
>
> $$B_{min} = 0.5 \times 1/T_s$$
>
> where T_s is the symbol period.

Knowing that the maximum symbol rate that can be supported on a channel is $2B$ symbols/second, and with each symbol conveying $\log_2 M$ bits, we can conclude that:

> The channel capacity for a *baseband channel* with bandwidth B Hz is:
>
> *Channel capacity* $C = 2B \log_2 M$ *bits/second*

Channel capacity restriction due to noise – the Shannon–Hartley theorem

As the number of symbol states M increases, the ability of the receiver to distinguish between symbols in the presence of noise and/or interference/

distortion decreases. Hence the ratio of *signal power S* to *noise power N* will be a crucial factor in determining how many symbol states can be utilized and still achieve 'error-free' communication.

The 'duration' of each symbol is also key in determining the noise tolerance of a receiver system, with longer symbols giving the receiver more time to average out the effects of noise than shorter symbols.

Noisy data pulse

Averaged data pulse

The combined effects of finite bandwidth *B* and finite signal to noise ratio *S/N* on channel capacity are governed by a very famous relationship known as the Shannon–Hartley capacity limit. The mathematical basis for this expression was first put forward in Shannon (1948a, 1948b).

> The Shannon–Hartley capacity limit for error-free communication is given by:
>
> $$\text{Channel capacity } C = B\log_2(S/N + 1) \text{ bits/second}$$

The Shannon–Hartley theorem states that if the required information transfer is less than the Shannon capacity limit (*C*), then error-free communication is possible. If information transfer at a rate greater than *C* is attempted, then errors in transmission will always occur no matter how well the equipment is designed.

The Shannon–Hartley capacity equation is a very good first step for evaluating the feasibility of any digital communication system design. It immediately provides the engineer with an 'upper bound' on channel capacity because it assumes a perfectly flat, distortion- and interference-free communications link, with the noise taking the form of Additive White Gaussian Noise (AWGN). It also is a theoretical bound with the implication that infinite signal processing power is available in both TX and RX units. In practice, we will of course not be able to achieve data rates quite as good as those suggested by the Shannon–Hartley equation, but it is a good starting point in a design.

E EXAMPLE 2.2

The specification for a Class 1 telephone link is a guaranteed flat bandwidth of 300 Hz to 3400 Hz and a minimum signal to noise ratio of 40 dB. The specification for a Class 2 telephone link is a guaranteed flat bandwidth from 600 Hz to 2800 Hz and a minimum signal to noise ratio of 30 dB. A company has a requirement to send data over a telephone link at a bit rate of 20 kbps without error. Would you advise the company to rent the more expensive Class 1 service or the cheaper Class 2 service? Justify your decision.

Solution

The Shannon–Hartley equation gives us the required relationship between channel capacity in bps, bandwidth and signal to noise ratio as follows:

[handwritten margin notes:] $10 \log(10,000) = 40\,dB$; $80\,dB$; $pur \leftarrow$; $20 \log_{volt}$

Channel capacity $C = B\log_2(S/N + 1)$ bps

For the Class 1 line, $B = 3400 - 300 = 3100\,$Hz and $S/N = 40\,$dB, thus,

$$C = 3100\log_2(10\,000 + 1) = 41.2\,\text{kbps}$$

Note, it is essential to convert the S/N value from dB to a ratio for use in the Shannon–Hartley expression.

For the Class 2 line, $B = 2800 - 600 = 2200\,$Hz and $S/N = 30\,$dB, thus

$$C = 2200\log_2(1000 + 1)\,\text{bps} = 21.9\,\text{kbps}$$

These calculations show that both the Class 1 and Class 2 lines will meet the specification of 20 kbps error-free transmission; however, the performance of the Class 2 line is very close to the Shannon bound, and allows little margin for error. In practice, it is unlikely that a modem could be realized that would give the desired result on the Class 2 line.

Power and bandwidth efficiency

For a system transmitting at maximum capacity, C, the *average signal power*, S, measured at the receiver input, can be written as $S = E_b \cdot C$, where E_b is the *average received energy per bit*.

The *average noise power*, N, can also be redefined as $N = N_0 \cdot B$, where N_0 is the *noise power density (Watts/Hz)*.

Using these definitions, the Shannon–Hartley theorem can be written in the form

$$C/B = \log_2(1 + E_b \cdot C/N_0 \cdot B)$$

The ratio C/B represents the *bandwidth efficiency* of the system in bits/second/Hz. The larger the ratio, the greater the bandwidth efficiency. The ratio E_b/N_0 is a measure of the *power efficiency* of the system. The smaller the ratio, the less energy used by each bit (and consequently for each symbol) to be detected successfully in the presence of a given amount of noise.

Choosing a power-efficient modem type is particularly important in cellular handsets, for example, where the designer is trying to maximize battery lifetime.

Power efficiency is covered in detail in Chapter 5 onwards.

Graphical representation

The Shannon–Hartley theorem clearly shows that bandwidth efficiency can be traded for power efficiency, and vice versa.

It is important to note that the Shannon capacity theorem assumes that the noise present with the signal is Additive White Gaussian Noise. This is often a valid assumption, particularly if the operating bandwidth is small compared with the channel centre frequency.

In practice, no implementation of a digital communications system can reach the performance suggested by the Shannon equation, and most fall short by 3 dB or more. Results for some common modem types are given in Chapter 6.

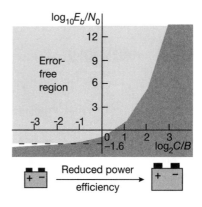

E EXAMPLE 2.3

A digital cellular telephone system is required to work at a bandwidth efficiency of 4 bits/second/Hz in order to accommodate sufficient users to make it profitable. What is the minimum E_b/N_0 ratio that must be planned for in order to ensure that users on the edge of the coverage area receive error-free communication?

If the operator wishes to double the number of users on his existing network, how much more power must the base-station and handsets radiate in order to maintain coverage and error-free communication?

Solution

The Shannon–Hartley theorem can be written as:

$$C/B = \log_2(1 + E_b \cdot C/N_0 \cdot B)$$

Now, the bandwidth efficiency is required to be $C/B = 4$ bits/second/Hz, thus:

$$4 = \log_2(1 + 4E_b/N_0)$$

therefore,

$$E_b/N_0 = (2^4 - 1)/4 = 3.75 \text{ or } 5.74 \text{ dB}$$

In order to double the number of users for the same operating bandwidth, the bandwidth efficiency of the system must be increased to 8 bits/second/Hz. This means that the E_b/N_0 value must rise to:

$$E_b/N_0 = (2^8 - 1)/8 = 31.87 \text{ or } 15.03 \text{ dB}$$

Thus the transmitted power must increase by a factor of $15.03 - 5.74 = 9.29$ dB.

QUESTIONS

2.1 A data link sends information in packets at a rate of 100 bits in 2.2 ms.

(a) What is the information rate supported by the channel during the packet burst?

(b) If the packets can only be sent every 5 ms, what is the overall information rate for the channel?

2.2 If the information capacity of a channel is 2400 bps, how long will it take to transfer 1 Mbyte of information between two computers?

2.3 A communications system represents four bits by each transmitted symbol. If the system is required to deliver a channel capacity of 9600 bps, what symbol rate must the channel be able to support?

2.4 If the symbol period as measured on a transmission cable is seen to be 2.5 ms, and the system specification states that each symbol represents six information bits, what is the channel capacity?

2.5 A mobile radio can support a data rate of 28 000 bps within a bandwidth of 25 kHz by encoding two bits in each symbol. What is the bandwidth efficiency of the radio link and what is the baud rate on the channel?

2.6 A data modem transfers information at 56 kbps, using 128 signalling states. What is the symbol rate for this application?

2.7 If a radio link is required to send digital voice at 4800 bps, but can only support a symbol rate of 1200 baud, how many symbol states must be used for this implementation?

2.8 A telephone transmission link has a usable channel bandwidth extending from 0 Hz to 3.1 kHz and can be assumed to be perfectly flat and distortion free. It is required to send information at a rate of 28.8 kbps over this channel.

What is the minimum number of symbol states that would be required to support this data rate?

2.9 If a 64 symbol state modem is designed to transfer data at a rate of 2.048 Mbps, what is the minimum bandwidth for the transmission cable, assuming baseband signalling?

2.10 A modem is to be designed for use over a telephone link, for which the available channel bandwidth is 3 kHz, and the average signal to noise ratio on the channel is 30 dB.

What is the maximum error-free data rate that can be supported on this channel and how many signalling states must be used?

2.11 A digital television transmission system must support a data rate of 3.5 Mbps within a bandwidth of less than 1.4 MHz. What is the maximum S/N ratio in dB that can be tolerated if the link is to provide error-free data communication?

2.12 An underwater communications link suffers from very high signal loss over a short distance, such that the maximum E_b/N_0 value achievable at the required range is only −0.6 dB. What is the maximum bandwidth efficiency that could be expected for this link at the extreme of range, and what is the information throughput that can be delivered in a bandwidth of 3400 Hz?

2.13 The serial ports on two computers which use binary signalling are connected by a twisted pair cable. The cable has a flat frequency response up to 12 kHz, with negligible group delay distortion.

(a) What is the maximum information transfer rate that can be accommodated by the cable, assuming a noise-free environment?

(b) If the noise introduced by the cable is −40 dB with respect to the signal power, what is the resulting maximum information transfer rate?

3 Baseband data transmission

3.1 Introduction

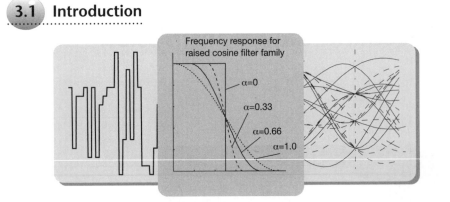

Frequency response for raised cosine filter family

Many people's concept of digital communications is one of nice square voltage pulses representing 0s and 1s being passed over a piece of cable or perhaps a radio system. In actual fact, this is rarely how digital information is sent. We have already seen the justification for using multi-level symbols to reduce the bandwidth content of a signal, and in this chapter we will see that square pulses become rounded pulses when passed through a channel with finite bandwidth.

In Section 3.2, the filtering effects of a channel are analysed for their impact on streams of data pulses – bits and symbols. What at first sight appears to be a 'show stopper' for data transmission – the intersymbol interference caused by channel filtering – is shown to be unfounded if steps are taken to achieve a Nyquist filter response for the whole system.

Section 3.3 introduces the eye diagram as a powerful visual tool for observing and diagnosing problems within the modem portion of a digital communications link. This is followed in Section 3.4 by a look at a very popular type of channel filter, the raised cosine filter, which is found in almost all modern modem implementations.

Section 3.5 discusses the concept of a matched filter – this in fact describes an overall channel filter response that should result in the optimum performance of the modem in the presence of noise. Finally, Section 3.6 goes against the stance taken in Section 3.3 by advocating the introduction of a controlled amount of intersymbol interference into the system by careful choice of filter (termed partial response signalling). Surprisingly, perhaps, this can achieve improved performance under certain conditions.

3.2 Intersymbol interference (ISI)

The problem of intersymbol interference

With any practical channel, the inevitable filtering effect will cause a spreading (smearing) of individual data symbols passing through the channel. For consecutive symbols, this spreading causes part of the symbol energy to overlap with neighbouring symbols, causing *intersymbol interference (ISI)*. Additionally, filtering in the transmitter or receiver units themselves may also introduce ISI degradation.

Unless very careful design steps are taken, intersymbol interference can *significantly degrade* the ability of the data detector to differentiate a current symbol from diffused energy of adjacent symbols. Even with no noise present in the channel this can lead to detection errors, termed an *irreducible error rate*, and at the very least will degrade the bit and symbol error rate performance in the presence of noise.

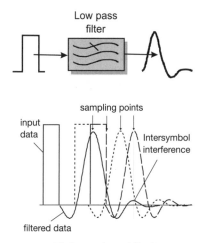

ISI due to channel filtering

Pulse shaping for zero ISI: Nyquist channel filtering

By carefully manipulating the filtering characteristics of the channel (and any TX or RX processing), it is possible to control the intersymbol interference such that it *does not degrade* the bit error rate performance of the link. This is achieved by making sure that the overall channel filter transfer function has what is termed a *Nyquist* frequency response.

A Nyquist channel response is characterized by the transfer function having a *transition band between passband and stopband that is symmetrical about a frequency equal to* $0.5 \times 1/T_s$.

For this type of channel response, the data symbols are still smeared, but the waveform *passes through zero* at multiples of the symbol period.

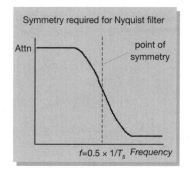

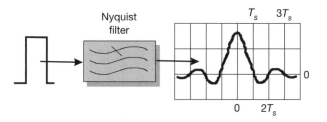

By sampling the symbol stream at the precise point where the ISI goes through zero, it can be seen that the energy spreading from adjacent symbols does not affect the value of the present symbol at that sampling point. It is also evident that the *sample timing must be very accurate* to minimize the ISI problem.

One of the major challenges in modem design, particularly in noisy or high distortion links, is the recovery of accurate symbol timing information (see page 59). Inaccuracy in symbol timing is usually referred to as *timing jitter*.

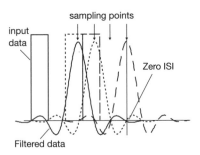

Zero ISI with raised cosine filters

Achieving a Nyquist channel response

It is *very unlikely* that the communications channel will inherently exhibit a Nyquist transfer response. This means that the system designer must add compensating filtering to achieve the desired response.

A telephone line modem presents a significant challenge for a zero ISI design as the transmission channel (see Section 4.4) itself may well introduce significant filter distortion which will compromise any carefully engineered Nyquist

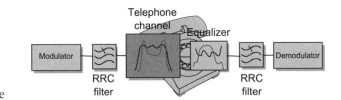

response between the sending and receiving hardware. In this scenario, it is necessary to try to correct for the channel distortion, or reconfigure the TX/RX filters so that the overall response is Nyquist.

Usually, adaptive channel equalizers are employed which attempt to flatten the channel transfer function so that the symbol shaping in the sending and receiving units predominates. For zero ISI performance, it is necessary to make sure that the *overall transfer function of these compound filters is Nyquist*. Most modern telephone modems operating at speeds above 4.8 kbps employ adaptive equalization using a 'training sequence' in the handshaking at the start of each call to measure the non-ideal channel response.

A good text on the subject of adaptive filters and equalizers is by Ifeachor and Jervis (1993).

Nyquist filtering – Example: cellular radio application

In some digital communications applications, digital radio for example, the transmission channel itself (the ether) may not impose any significant filtering effect across the modulation bandwidth, and the main filtering is performed by transmitter and receiver circuitry.

This transmitter filtering is employed largely to constrain the modulation to the regulated bandwidth, while in the receiver, the filtering is necessary to remove a multitude of other signals entering the receiver, and to minimize the noise entering the demodulator. Often the Nyquist filtering response needed for zero ISI is split equally between the TX and RX systems using a root raised cosine filter pair (see Section 3.4).

3.3 Eye diagrams

Generation of eye diagrams

The *eye diagram* is a convenient visual method of diagnosing problems with data systems.

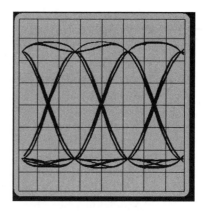

An eye diagram is generated conventionally using an oscilloscope connected to the demodulated, filtered symbol stream, *prior to* conversion of the symbols to binary digits. The oscilloscope is re-triggered at every symbol period or fixed multiple of symbol periods using a symbol timing signal (see Section 3.4) derived from the received waveform. By relying on the persistence of a typical oscilloscope display, the result is an overlaying of consecutive received symbol samples which form an 'eye' pattern on the screen. (This effect is re-created using modern digital storage scopes or computer displays in test equipment.)

The eye diagram is used throughout the remainder of this book to illustrate the effect of differing channel responses and source of degradation on the data receiver performance.

Diagnosis using the eye diagram

From the eye diagram it is possible to make an engineering judgement on the likely performance and sources of degradation in a data communications link.

Shown here are examples of eye diagrams for various types of distortion – each having a unique identifiable effect on the appearance of the 'eye opening'.

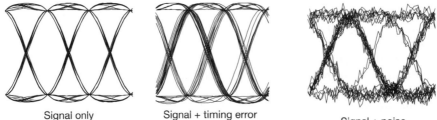

Signal only   Signal + timing error   Signal + noise

The effect of timing error is seen as a skewing of the eye diagram and a closing of the eye due to the received symbol stream no longer being sampled at the point of zero ISI. The addition of noise affects the timing recovery circuitry and also causes a general closing of the eye until eventually the noise occasionally causes full 'eye closure' and errors occur.

Example of complex eye diagram

Shown here are the eye diagrams for a modulation scheme with four states and 16 states in the demodulated signal. This would be typical for one branch of a 16-QAM (Quadrature Amplitude Modulation) and 256-QAM modem. The eye diagram as a diagnosis tool comes into its own here, showing clearly the individual 'eyes' between the discrete states, and also illustrating how critical the sample timing must be to detect the symbol at the maximum eye opening.

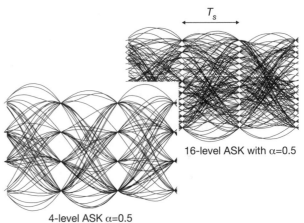

16-level ASK with α=0.5

4-level ASK α=0.5

It also serves to show how much more susceptible high-order modulation formats are to noise and distortion when compared with a binary system of equivalent energy per bit. The eye opening for the four-level system is so much narrower and the spacing between decision boundaries so much smaller than those in the previous two-state examples and this gets progressively worse as the number of symbol states rises.

Great care should be exercised when using the eye diagram for diagnosis to ensure that the observation is made after all the filtering within the system.

3.4 Raised cosine filtering

Raised cosine filter family

A commonly used realization of the Nyquist filter is a *raised cosine filter*, so called because the transition band (the zone between passband and stopband) is shaped like part of a *cosine* waveform.

The sharpness of the filter is controlled by the parameter α, *the filter roll-off factor*. When $\alpha = 0$ this conforms to an ideal brick-wall filter.

The bandwidth B occupied by a raised cosine filtered data signal is thus *increased* from its minimum value, $B_{min} = 0.5 \times 1/T_s$, to:

Actual modulation bandwidth,
$$B = B_{min}(1 + \alpha)$$

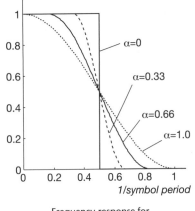

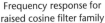

Frequency response for raised cosine filter family

E EXAMPLE 3.1

A four-level baseband data stream has a symbol period of 100 μs.

(a) What is the minimum bandwidth required for transmission, assuming a root raised cosine filter is used with $\alpha = 0.3$?

(b) What is the time taken to transmit 1 million bits?

(c) If it is required to transmit the information in half the time, how many symbol states would be required using the same transmission bandwidth?

Solution

(a) The minimum bandwidth required for transmission is half the symbol period for baseband signalling, assuming brick-wall filtering ($\alpha = 0$). For $\alpha = 0.3$, this bandwidth must be increased by a factor of $(1 + \alpha)$. Thus for a symbol period of 100 μs, the symbol rate is 10 000 symbols/second, and the bandwidth required is $5000(1 + \alpha) = 65\,000$ Hz.

(b) With a four-level signalling scheme, there can be two bits encoded in each symbol, hence the bit rate is twice the symbol rate, i.e. 20 kbps. The time taken to send 1 million bits is thus $1\,000\,000/20\,000 = 50$ seconds.

(c) To halve the transmission time, we need to encode double the number of bits in each symbol. Thus 4 bits/symbol requires 16 symbol states.

Impulse response of raised cosine filter

The spreading (ISI) effect of a raised cosine filter on the data pulses passing through it can be found by plotting the *impulse response* of the filter.

MAT LAB

The amount of 'ringing' produced by the filter is dependent on the α chosen. The smaller the value of α, the more pronounced the ringing as the filter response approaches that of the ideal brick-wall filter ($\alpha = 0$).

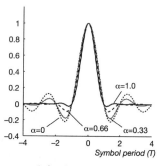

Pulse time response for raised cosine filter family

IN DEPTH

Transfer function and impulse response of a raised cosine filter

The raised cosine filter is one of the family of Nyquist filters that satisfy the criteria for zero intersymbol interference.

The transfer function $H(f)$ of a raised cosine filter is given by:

$$H(f)_{\text{raised-cosine}} = \begin{cases} T & \text{for } 0 \leqslant |f| \leqslant \dfrac{1-\alpha}{2T} \\[2ex] \dfrac{T}{2}\left[1 + \cos\left(\dfrac{\pi T}{\alpha}\left\{|f| - \dfrac{1-\alpha}{2T}\right\}\right)\right] & \text{for } \dfrac{1-\alpha}{2T} \leqslant |f| \leqslant \dfrac{1+\alpha}{2T} \\[2ex] 0 & \text{for } \dfrac{1+\alpha}{2T} \leqslant |f| \end{cases}$$

where α is the filter roll-off factor and T is the symbol period of the message. An $\alpha = 0$ corresponds to an ideal brick-wall filter.

The impulse response for a raised cosine filter is given by:

$$h(t) = \frac{\text{sinc}(t/T)\cos(\pi\alpha t/T)}{1 - 4\alpha^2 t^2/T^2}$$

Eye diagrams for raised cosine filtered data

The effect of changing α on the time domain response of the data signal is readily seen from the eye diagram. As α is reduced, the eye opening dramatically narrows, requiring the accuracy of symbol timing (see page 59) to be even more exact. Additionally, it can be seen that the 'overshoot' of the pulse caused by the filtering is greater for small α, increasing the peak to mean ratio of the energy in the data signal, and hence the peak signal handling requirement of the modulator and demodulator circuits.

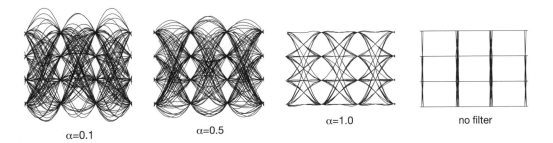

α=0.1 α=0.5 α=1.0 no filter

The *root* raised cosine filter

Given that the Nyquist filter response must apply to the *whole transmission link* including TX and RX sections to achieve zero ISI performance, it is necessary to consider where best to place the filtering within the transmission system.

Filtering is almost always mandatory in the transmission unit, particularly in the case of wireless communications, to constrain the 'on-air bandwidth' of the signal to that dictated by regulation (see Section 2.1) or by the practical necessity to co-habit with users on adjacent channel frequencies. Good receiver filtering is also vital (again especially in wireless applications), in order to remove strong signal interferers from overloading the demodulator circuitry, and also to reject as much noise as possible that does not fall within the modulation passband. For these reasons, it is necessary for the Nyquist filter function to be shared between TX and RX units (assuming that the channel response itself is flat or has been equalized (see Section 4.5) in some way).

It is common practice to split the filtering function of a raised cosine filter equally between transmitter and receiver so that each unit ends up with what is termed a *root raised cosine* transfer

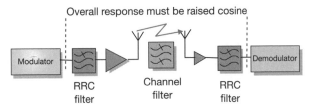

characteristic. Under these circumstances, the peak to mean ratio of the transmitted signal must be evaluated for a *root* raised cosine response, not a raised cosine response as is often mistakenly thought.

Implementation of digital Nyquist filters

Traditionally it has been difficult to construct a filter having a Nyquist response using analogue components, and it has taken the development of the digital signal processor (DSP) (see Section 2.1) to bring Nyquist and raised cosine filters into everyday use.

Using a class of filters known as *digital non-recursive linear phase* or *finite impulse response (FIR)* filters, it is possible to approximate to any required degree of accuracy a perfect raised cosine filter. The 'length' of the filter (equivalent to

IN DEPTH

Impulse response of a *root* raised cosine filter

The root raised cosine filter is so called because the transfer function is exactly the square root of the transfer function of the raised cosine filter. Hence:

$$
H(f)_{\text{root-raised-cosine}} = \begin{cases} \sqrt{T} & \text{for } 0 \leqslant |f| \leqslant \dfrac{1-\alpha}{2T} \\[2ex] \sqrt{\dfrac{T}{2}\left[1 + \cos\left(\dfrac{\pi T}{\alpha}\left\{|f| - \dfrac{1-\alpha}{2T}\right\}\right)\right]} & \text{for } \dfrac{1-\alpha}{2T} \leqslant |f| \leqslant \dfrac{1+\alpha}{2T} \\[2ex] 0 & \text{for } \dfrac{1+\alpha}{2T} \leqslant |f| \end{cases}
$$

where α is the filter roll-off factor and T is the symbol period of the message. An $\alpha = 0$ corresponds to an ideal brick-wall filter.

The impulse response for a root raised cosine filter is given by:

$$
h(t) = \frac{4\alpha}{\pi\sqrt{T}} \frac{\cos[(1+\alpha)\pi t/T] + \dfrac{T}{4\alpha t}\sin[(1-\alpha)\pi t/T]}{1 - (4\alpha t/T)^2}
$$

filter order in analogue terminology) is greater for filters that have sharp transition bands, corresponding to filters with small α.

Long filters incur the greatest processing overhead and also introduce the largest propagation delay through the filter. High filter delay is particularly disadvantageous in applications such as full-duplex speech transmission, where the subjective effect of time delay in a conversation

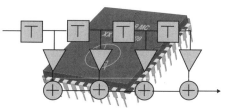

Digital filter structure for implementing RC filter

can be very disturbing. (Note: significant delay is also introduced by the speech vocoders and other aspects of the data modulation and demodulation process such as interleaving and error correction (see Chapter 7).)

Care with eye diagrams

Eye diagrams (Section 3.3) are a very powerful tool for visual diagnosis of the degree of distortion and noise introduced by the channel (including TX and RX processing) and of any errors in symbol timing recovery.

Often, however, it is not possible to observe the eye diagram at the *correct*

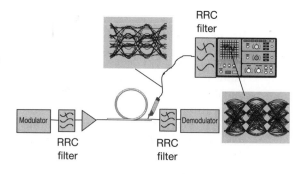

point in the circuit, namely after any receiver filtering where the full (hopefully) Nyquist response should be seen, as this may be buried within an IC implementation.

Under these circumstances it is necessary to *duplicate* any filtering following the observation point within the test hardware itself so that a representative eye diagram is obtained.

Symbol timing recovery

The eye diagram is a very graphic way of highlighting the need for accurate symbol timing in the receiver in order to sample the received signal at the maximum eye opening.

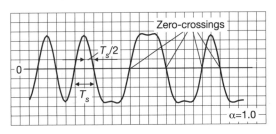

Symbol timing can be derived by sending a timing reference alongside the data signal, for example in the form of a continuous tone at a multiple or sub-multiple of the symbol rate (Bateman, 1990), or in the form of a burst clock between message data transmissions. Most symbol timing systems, however, obtain their information from the message data itself, making use of the 'zero-crossing' information in the baseband (bipolar (see Section 1.4)) signal.

The problem of symbol timing is eased significantly for raised cosine filtered signals if the roll-off factor α is equal to 1. For this unique case, the zero-crossings of the filtered waveform always occur at a time $T_s/2$ before the optimum zero ISI (see Section 3.2) detection point. Triggering a timer from these zero-crossings to sample the signal $T_s/2$ seconds later will thus give perfect symbol timing. When the data contains long strings of 1s or 0s, the sampling process must interpolate the correct sample times until the next zero-crossing occurs. Data 'scrambling' can be used to try to increase the frequency of zero-crossings in the data stream.

Symbol timing circuits

The main drawback with using an $\alpha = 1$ data system to achieve symbol timing is the sacrifice in bandwidth that such a large α entails. Also, when the received signal is accompanied by noise, individual zero-crossings may be corrupted by the noise and some form of averaging is desirable over many zero-crossings if accurate timing is to be achieved.

One circuit that can average the effects of noise as well as the non-perfect zero-crossings caused by smaller values of α is the feedback timing control loop shown below. The circuit operates by using a mono-stable to create pulses of duration $T_s/2$ at each zero-crossing, which are then compared in a digital

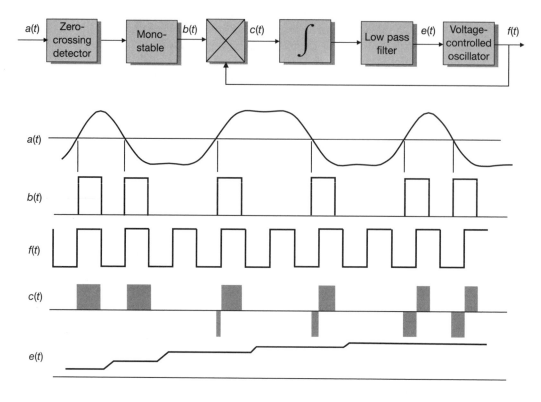

mixer with a locally generated clock running close to the symbol rate. The output of the mixer is integrated and filtered in order to produce a smoothed dc control voltage that is used to nudge the clock on to the correct symbol rate. The loop filtering process also serves to average out the noise accompanying the input signal and zero-crossing variations due to the small values of α. In practice, there is a compromise between wishing to quickly acquire the symbol timing for rapid data decoding, and having a long averaging time to minimize 'timing jitter'. Two excellent references on timing and carrier recovery circuits are Lindsey and Simon (1972) and Gardner (1966).

An alternative symbol timing recovery circuit involves 'squaring' the received filtered data stream, which yields a signal with a strong discrete frequency component at the symbol timing frequency. Extracting this signal with a

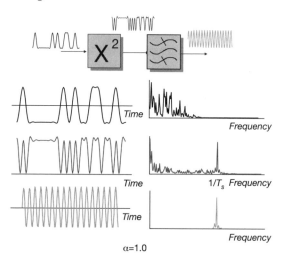

narrow tuned filter or a phase-locked loop (see Section 5.3) gives the required symbol clock. This system works well for $\alpha = 1$, but will not produce a discrete spectral line for small values of α.

Yet a third method of symbol timing recovery is the early-late gate method. This technique is described in connection with Amplitude Shift Keying in Chapter 5.

Summary: choice of α

Benefits of small α

- Maximum bandwidth efficiency achieved.

Benefits of large α

- Simpler filter – fewer stages (taps) hence easier to implement with less processing delay.
- Less signal overshoot, resulting in lower peak to mean excursions of the transmitted signal.
- Less sensitivity to symbol timing accuracy – wider eye opening.

3.5 Matched filtering

Recovery of symbols from noise

Having dealt earlier with the issue of controlling the bandwidth of a data signal and filtering for minimum intersymbol interference (see Section 3.2), we shall now turn our attention to the question of how to optimize the detection of data symbols in the presence of

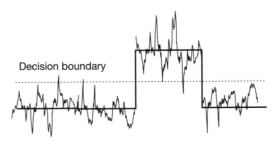

noise. This discussion will be limited to the case of Additive White Gaussian Noise, which simplifies the mathematical treatment considerably. For many applications, this fortunately represents a good approximation to reality.

Consider a binary symbol stream accompanied by noise entering the receiver. At some point during each symbol the receiver has to make a decision as to whether a 1 or 0 has been sent. The logical decision point is thus halfway between the two voltage levels representing each symbol.

The chances of the receiver making the correct decision are of course greatly improved if the detector incorporates some form of 'averaging' (filtering) in order to improve the signal to noise ratio at a particular sampling point.

For example, passing the symbol through an integrator would gradually improve the S/N ratio, reaching a maximum at the end of the symbol period. If the integrator continued to average over the following symbol (in this case a zero), the output would begin to fall and would yield a non-optimum result for both the 1 and 0 symbols. For this reason it would be prudent to reset the integrator at the beginning of each new symbol. This type of detector is commonly described as an *integrate and dump* filter.

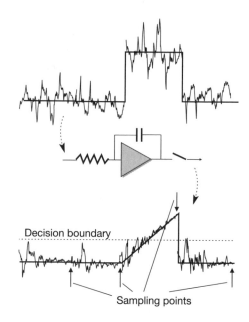

The concept of an optimum (matched) filter

It is prudent to ask the question: 'What type of averaging filter will give the best S/N ratio at the sampling point?' The answer is that it depends entirely on the symbol shape being used.

Consider the two symbol shapes shown here, one a square (unshaped pulse) and the other a rounded pulse. For the square pulse, the S/N ratio at each point in the symbol is approximately constant, and an averaging process that gives equal weight to each point would be optimum. This in fact is exactly what an integrate and dump filter does. For the rounded pulse, however, it is evident that the S/N ratio is greater in the centre of the pulse than at the edges, and it would thus make sense to give more weight to averaging the central region rather than the edges of the symbol. Hence the integrate and dump filter would not be optimum for this shape of symbol. A detection filter that does optimize the S/N ratio for a symbol is called a *matched filter* because its averaging effect is matched to the pulse shape. The integrate and dump filter is thus a matched filter for a rectangular pulse shape, but would not be matched to a root raised cosine pulse, for example.

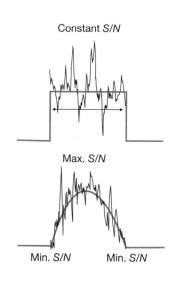

Designing a matched filter with ISI in mind

It can be shown (see in-depth section) that a detection filter will be 'matched' to the input symbol pulse shape if the filter is designed with an impulse response that is in fact a time reversed and delayed replica of the input symbol shape. Alternatively, the frequency domain response of a matched filter must be equal to the complex conjugate of the spectrum of the input symbol. Clearly, implementing a matched filter requires detailed knowledge of the source data symbol shape, and also relies on the symbol shape remaining undistorted as it passes through the channel. It also only holds true for Additive White Gaussian Noise.

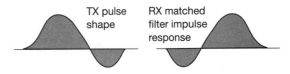

TX pulse shape   RX matched filter impulse response

    If we can assume that the channel is distortion free or has been equalized (see Section 4.5) to remove the distortion, then we can introduce the concept of a *matched filter pair*, with one filter performing the pulse shaping in the transmitter and the other performing matched detection in the receiver. We already have a requirement for paired filtering, however, driven by the desire to achieve zero intersymbol interference. The important question is, does the Nyquist filtering requirement for zero ISI (see Section 3.2) conflict with the matched filtering requirement for optimum S/N ratio? Fortunately it has been proven mathematically (Schwartz, 1990) that the root raised cosine filter (Section 3.4) pair satisfies both criteria, which explains its popularity with modem designers around the world.

IN DEPTH

Criteria for matched filtering in AWGN

Consider the response of a signal $s(t)$ plus noise $n(t)$ passing through a detection filter with transfer function $H(f)$. If the Fourier transform of the signal is $S(f)$ then the time domain output of the filter $s_0(t)$ due to the signal component alone is given by:

$$s_0(t) = \int_{-\infty}^{\infty} H(f)S(f)e^{j2\pi ft} dt$$

and the output signal power S is proportional to the square of the signal voltage, thus:

$$S = |s_0(t)|^2 = \left| \int_{-\infty}^{\infty} H(f)S(f)e^{j2\pi ft} dt \right|^2$$

    The power spectral density of the noise at the filter output $N_0(f)$ is given by the squared magnitude of the filter transfer function multiplied by the power spectral density of the input noise.

For AWGN, we know the noise spectral density is flat with a value N_0 Watts/Hz and hence the output noise spectral density is:

$$N_0(f) = N_0 |H(f)|^2$$

The average noise power N is found by integrating the noise power density over all possible frequencies to give:

$$N = N_0 \int_{-\infty}^{\infty} |H(f)|^2 df$$

The goal of a matched filter is to make the signal to noise ratio at the sampling time $t = T$ a maximum. A matched filter will thus need to optimize the S/N ratio given by:

$$S/N = \frac{\left| \int_{-\infty}^{\infty} H(f) S(f) e^{j2\pi fT} dt \right|^2}{N_0 \int_{-\infty}^{\infty} |H(f)|^2 df}$$

In order to find the transfer function $H(f)$ which maximizes the S/N ratio we need to make use of a result known as Schwarz's inequality.

Schwarz's inequality states that:

$$\left| \int_{-\infty}^{\infty} x(t) y(t) dt \right|^2 \leqslant \int_{-\infty}^{\infty} x^2(t) dt \int_{-\infty}^{\infty} y^2(t) dt$$

and also states that for the two sides of this expression to be equal, then:

$$X(f) = Y^*(f) e^{-j2\pi fT}$$

Applying this relationship to the S/N expression for the filter output we get:

$$S/N \leqslant \frac{1}{N_0} \int_{-\infty}^{\infty} |S(f)|^2 df$$

where we have made use of the fact that $|e^{j2\pi fT}| = 1$.

The value of the S/N ratio is maximized when this expression is equal. The Schwarz inequality thus allows us to conclude that for optimum S/N ratio, that is, a matched filter, then:

$$H_{\text{matched}}(f) = S^*(f) e^{-j2\pi fT}$$

The impulse response for this matched filter is thus given by:

$$h_{\text{matched}}(t) = \int_{-\infty}^{\infty} S^*(f) e^{-j2\pi f(t-T)} df$$

Given that $S^*(f) = S(-f)$ for a real-valued signal $s(t)$, then:

$$h_{\text{matched}}(t) = \int_{-\infty}^{\infty} S(-f)e^{-j2\pi f(t-T)}df = s(T-t)$$

This result tells us that the impulse response of a matched filter should be a time reversed and delayed version of the input symbol $s(t)$.

Bit error rate (BER) performance for baseband data systems

In order to determine the likelihood of a data detector decoding symbols correctly in the presence of noise, we have to operate with probabilities and statistics. Noise is *non-deterministic*, that is, its amplitude and phase vary randomly with time so that during some symbols the instantaneous levels of noise will be greater than for others. We thus must

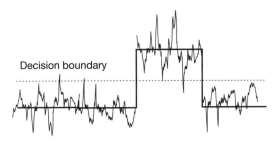

resign ourselves to working out the *probability* of decoding symbols in error, rather than the certainty of decoding symbols in error.

Often in fact we wish to be informed on the probability of bit error rather than symbol error as this directly impinges on the integrity of data sent to the user. For the binary case, symbol error probability and bit error probability are of course identical.

BER performance for matched filter detection

Let us consider the BER performance for a rectangular baseband pulse using an integrate and dump filter for detection. (We know that this is a matched filter detector for this pulse shape and hence we can infer that this result will apply for all matched filter detectors regardless of the pulse shape.)

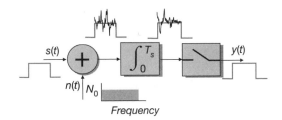

If we look at a single symbol $s(t)$ of voltage V passing through the detector, with additive noise $n(t)$, then the output of the integrator $y(t)$ can be written as:

$$y(t) = \int_0^T \{s(t) + n(t)\}dt = V \cdot T + \int_0^T n(t)dt$$

The contribution at the sampling instant due to the symbol is thus a voltage $V \cdot T$ volts. The symbol will be detected correctly if the contribution due to noise does not take the signal plus noise sample below the decision threshold ($V \cdot T/2$ volts) which is halfway between 0 volts for the logic 0 case, and $V \cdot T$ volts for the logic 1 case.

The statistics of the integrated noise at the output of the integrate and dump detector are such that the probability density of the noise samples follows a Gaussian distribution as shown here. The probability of making a decision error is thus the probability of the noise sample being more negative than $-V \cdot T/2$ volts.

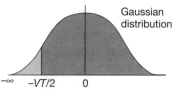

Gaussian distribution

$-\infty$ $-VT/2$ 0

The *probability of symbol error* P_s (see in-depth section) is therefore given by the expression:

$$P_s \text{ (symbol error probability)} = (0.5 \cdot erfc)(0.25 \cdot E_s/N_0)^{1/2}$$

where E_s is the energy in the logic 1 symbol, and N_0 is the noise power density.

If the symbol to be detected were a logic 0 (that is, 0 volts) rather than a logic 1, then exactly the same reasoning can be applied, except we need to evaluate the probability that the noise sample produces a positive value greater than $+V \cdot T/2$ volts for an error to occur. Owing to the symmetry of the Gaussian distribution about $0\,V$, this probability of error is identical to that for the logic 1 case.

IN DEPTH

Probability of symbol error for binary baseband unipolar data detection

For an input symbol of voltage V, integrated over a period T seconds, the output voltage will be $V \cdot T$ volts.

For AWGN, the effect of integrating the noise waveform over a period T is to create a noise output voltage which has a Gaussian distribution with variance $\sigma_o^2 = N_0 T/2$, where N_0 the the noise power density in Watts/Hz.

The probability density of the integrated noise, $n_o(t)$, at the sampling point is thus:

$$P_d[n_o(t)] = \frac{e^{-n_o^2(T)/2\sigma_o^2}}{\sqrt{2\pi\sigma_o^2}} = \frac{e^{-n_o^2(T)\cdot N_0 T}}{\sqrt{\pi N_0 T}}$$

A detection error will occur when the noise sample exceeds a value $-VT/2$. The probability of this occurring will thus equal the symbol error probability and is given by:

$$P_s = \int_{-\infty}^{-VT/2} P_d[n_o(t)]\,dn_o(t) = \int_{-\infty}^{-VT/2} \frac{e^{-n_o{}^2(T)\cdot N_0 T}}{\sqrt{\pi N_0 T}}\,dn_o(t)$$

Defining $x = n_o(T)/\sqrt{N_0 T}$, the error probability can be written as:

$$P_s = \frac{1}{2}\frac{2}{\sqrt{\pi}}\int_{x=-\infty}^{-\frac{V}{2}\sqrt{\frac{T}{N_0}}} e^{-x^2}\,dx = \frac{1}{2}erfc\left(\frac{V}{2}\sqrt{\frac{T}{N_0}}\right)$$

where $erfc(x)$ is known as the complementary error function and frequently appears in the analysis of digital communications systems.

We can express this symbol error probability in terms of the symbol energy as follows:

$$P_s = \frac{1}{2}erfc\left(\frac{V}{2}\sqrt{\frac{T}{N_0}}\right) = \frac{1}{2}erfc\left(\frac{1}{4}\frac{V^2 T}{N_0}\right)^{1/2} = \frac{1}{2}erfc\left(\frac{1}{4}\frac{E_s}{N_0}\right)^{1/2}$$

The probability of receiving a logic 0 symbol in error yields exactly the same result, although in this case we are looking for the probability of the noise sample exceeding $+VT/2$.

Recognizing that the average symbol energy for a unipolar data stream is half that of the symbol energy for the logic 1 symbol (assuming an equal likelihood of 1s and 0s being sent), the symbol error probability for unipolar baseband transmission can be written as:

$$P_{s_{unipolar}} = \frac{1}{2}erfc\left(\frac{1}{2}\frac{E_s}{N_0}\right)^{1/2}$$

Unipolar vs bipolar symbols

The energy per symbol for the *unipolar* waveform considered thus far is different depending on whether a logic 0 or logic 1 is sent, having a zero value for the logic 0 case. Assuming an equal likelihood of 1s and 0s being transmitted, the *average energy per symbol* sent is $E_s/2$ where E_s, as before, is the energy of the logic 1 symbol.

> The symbol error probability for a *unipolar* binary data system with matched filtering is thus:
>
> $$P_{s\,unipolar} = 0.5 \cdot erfc(0.5 \cdot E_{s_{average}}/N_0)^{1/2}$$

Unipolar: $P_s=0.5\ erfc[\sqrt{(E_s/2N_0)}]$
Bipolar: $P_s=0.5\ erfc[\sqrt{(E_s/N_0)}]$

If we now look at the *bipolar* data waveform where a logic 1 is conveyed as $+V$ volts and a logic 0 as $-V$ volts, we can show (see in-depth section) that:

The symbol error probability for a *bipolar* waveform is given by:

$$P_{s\,bipolar} = 0.5 \cdot erfc(E_{s_{average}}/N_0)^{1/2}$$

It is immediately apparent that the bipolar signalling method requires only half the average symbol energy for a given probability of error compared with the unipolar case.

IN DEPTH

Probability of symbol error for binary baseband bipolar data detection

For an input symbol of voltage V, integrated over a period T seconds, the output voltage will be $V \cdot T$ volts.

For AWGN, the effect of integrating the noise waveform over a period T is to create a noise output voltage which has a Gaussian distribution with variance $\sigma_o^2 = N_0 T/2$, where N_0 the the noise power density in Watts/Hz.

The probability density of the integrated noise, $n_o(t)$, at the sampling point is thus:

$$P_d[n_o(t)] = \frac{e^{-n_o^2(T)/2\sigma_o^2}}{\sqrt{2\pi\sigma_o^2}} = \frac{e^{-n_o^2(T)\cdot N_0 T}}{\sqrt{\pi N_0 T}}$$

For the bipolar signal, a detection error will occur when the noise sample exceeds a value $-V \cdot T$ and hence the probability of this occurring, which equals the symbol error probability, is given by:

$$P_s = \int_{-\infty}^{-VT} P_d[n_o(t)]dn_o(t) = \int_{-\infty}^{-VT} \frac{e^{-n_o^2(T)\cdot N_0 T}}{\sqrt{\pi N_0 T}} dn_o(t)$$

Defining $x = n_o(T)/\sqrt{N_0 T}$, the error probability, P_s, can be written as:

$$P_s = \frac{1}{2}\frac{2}{\sqrt{\pi}}\int_{x=-\infty}^{-V\sqrt{\frac{T}{N_0}}} e^{-x^2} dx = \frac{1}{2} erfc\left(V\sqrt{\frac{T}{N_0}}\right)$$

where $erfc(x)$ is known as the complementary error function and frequently appears in the analysis of digital communications systems.

We can express this symbol error probability in terms of the symbol energy as follows:

$$P_s = \frac{1}{2} erfc\left(V\sqrt{\frac{T}{N_0}}\right) = \frac{1}{2} erfc\left(\frac{V^2 T}{N_0}\right)^{1/2} = \frac{1}{2} erfc\left(\frac{E_s}{N_0}\right)^{1/2}$$

The probability of receiving a logic 0 symbol in error yields exactly the same result, where we are now looking for the probability of the noise sample exceeding $+V \cdot T$.

Recognizing that the average symbol energy for a bipolar data stream is the same for both a logic 0 and a logic 1 symbol, then the symbol error probability is unchanged and can be written as:

$$P_{S_{bipolar}} = \frac{1}{2} erfc\left(\frac{E_s}{N_0}\right)^{1/2}$$

BER performance for M-ary signalling

We would expect intuitively that as we increase the number of symbol states, the ability of the receiver to distinguish between symbols in the presence of noise will decrease, unless we significantly increase the energy in each symbol. Based on exactly the same type of analysis used for the binary case, it can be shown (Proakis, 1983) that:

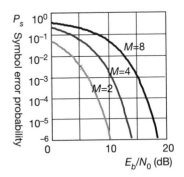

The *symbol error probability for M-ary bipolar baseband signalling* is given by:

$$P_{S_{M\text{-aryBipolar}}} = [(M-1)/M] \cdot erfc([3/(M^2-1)] \cdot E_{S_{average}}/N_0)^{1/2}$$

where M is the number of symbols used.

This result is plotted here as a function of the average energy per bit E_b rather than the average energy per symbol E_s where $E_b = E_s/k$ (k is the number of bits per symbol).

EXAMPLE 3.2

A company wishes to increase the throughput of a telephone modem product by changing from a two-level baseband signalling scheme to an eight-level signalling scheme and has set a design target of maintaining a performance of no worse than one symbol error in every 10 000 symbols sent. By using the plot of symbol error vs E_b/N_0 for M-ary baseband signalling, determine the reduction in noise tolerance for the modem as a result of this change.

What is the theoretical minimum E_b/N_0 required to support the bandwidth efficiency achievable by the eight-level modem?

Solution

From the plot of symbol error for M-ary signalling, at a symbol error probability of 1 in 10^4, it can be seen that an increase of about 8 dB in signal energy is required to maintain the same error rate. In other words, the new modem will be approximately 8 dB less tolerant to noise. At the 1 in 10^4 error rate level, the 8-ary modem requires approximately 12.5 dB E_b/N_0.

An 8-ary baseband modem has a maximum spectral efficiency of 6 bits/second/Hz.

The Shannon–Hartley theorem states that:

$$C/B = \log_2[1 + E_b \cdot C/N_0 \cdot B]$$

Thus:

$$6 = \log_2[1 + 6 \cdot E_b/N_0]$$

therefore the minimum E_b/N_0 for error-free transmission is:

$$E_b/N_{0min} = (2^6 - 1)/6 = 10.5 \text{ or } 10.2 \text{ dB}$$

Bit error rate vs symbol error rate

So far we have derived the probability of symbol error for a data system whereas in practice the user is more concerned with the probability of bit error for the communications link.

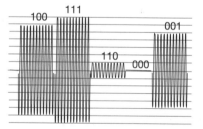

For a binary system, bit and symbol error are identical as each symbol error corresponds to a single bit error. For M-ary systems $(M > 2)$, however, this does not hold true. For example, a 16-ary scheme conveying four bits per symbol may incur anything from one bit error to four bit errors for each incorrectly decoded symbol, depending on which symbol was mistakenly identified.

In practice, some symbols are more likely to be detected in error than others, depending on how close or similar the symbols are to the correct symbol. Careful choice of bit pattern assignment to each symbol may thus help to minimize the number of bit errors occurring for every symbol error.

Gray coding

Gray coding is the name given to a bit assignment where the bit patterns in adjacent (most similar) symbols only differ by one bit. If we then make an assumption that the detection process will only mistake symbols for those adjacent to the correct symbol, we can infer that the bit error probability will be given approximately by the symbol error probability divided by the number of bits k in each symbol, that is:

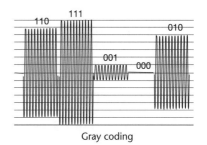

Gray coding

$$P_b(\text{bit error probability})_{\text{Gray coding}} \approx P_s(\text{symbol error probability})/k$$

3.6 Partial response signalling

Partial response signalling, or *correlative coding* as it is sometimes called, is a baseband signalling technique that *deliberately* introduces large amounts of intersymbol interference into the transmitted signal in order to ease the burden on the pulse-shaping filters.

The reasoning is that if intersymbol interference is added in a simple and controlled manner, then it should be possible to subtract out the interference in the receiver. Removing the need to strive for Nyquist filter (see Section 3.2) operation with minimum bandwidth, high roll-off filters is the motivation behind the partial response technique.

A whole family of partial response signalling methods exists; however, by way of introduction we will discuss only two very simple and well-known forms – *duo-binary* and *modified duo-binary* signalling.

Duo-binary signalling

In a duo-binary system, the input data sequence is combined with a 1-bit delayed version of the same sequence (the controlled intersymbol interference) and then passed through the pulse-shaping filter. A binary input with voltages $+V$ and $-V$ thus produces a three-level output with voltages $+2\,\text{V}$, $0\,\text{V}$ and $-2\,\text{V}$.

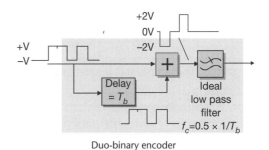

Duo-binary encoder

For simplicity, if we make the pulse-shaping filter an 'ideal brick-wall filter' with cut-off at $0.5 \times 1/T_b$, then the impulse response and frequency response of the combined duo-binary encoder and pulse-shaping filter are as shown (see in-depth section). As expected, energy from one input bit is seen to be spread over two bit periods in the transmitted signal. The composite frequency response, however, is the key to the success of duo-binary signalling. A filter with this response is relatively easy to synthesize and can be used to *replace* the duo-binary encoder and unrealizable brick-wall filter and *realistically* achieve a data throughput of 2 bits/second/Hz.

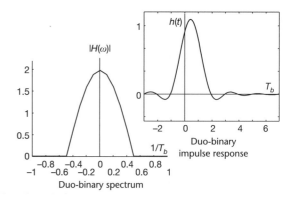

Duo-binary spectrum

Duo-binary impulse response

IN DEPTH

Calculation of impulse response and frequency response for duo-binary signalling

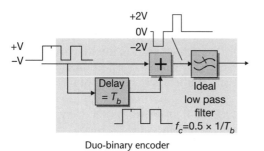

Duo-binary encoder

The transfer function of an ideal delay element T_b is $e^{-j2\pi f T_b}$. The transfer function of the ideal low pass filter is simply:

$$H(f)_{LP} = \begin{cases} 1 & \rightarrow |f| \leqslant 0.5 \times 1/T_b \\ 0 & \rightarrow \text{otherwise} \end{cases}$$

Thus the overall frequency response of the duo-binary scheme is:

$$\begin{aligned} H(f)_{Duobinary} &= H(f)_{LP}[1 + e^{-2j\pi f T_b}] \\ &= H(f)_{LP}[e^{j\pi f T_b} + e^{-j\pi f T_b}]\, e^{-j\pi f T_b} \\ &= H(f)_{LP}2 \cdot \cos(\pi f T_b)\, e^{-j\pi f T_b} \end{aligned}$$

$$H(t)_{\text{Duobinary}} = \begin{cases} 2\cos(\pi f T_b)\, e^{-j\pi f T_b} & \to |f| \leqslant 0.5 \times 1/T_b \\ 0 \end{cases}$$

The impulse response $h(t)$ for the duo-binary scheme is simply the sum of two *sinc waveforms*, delayed by one bit period with respect to each other:

$$h(t)_{\text{Duobinary}} = \frac{\sin(\pi t/T_b)}{\pi t/T_b} + \frac{\sin(\pi(t - T_b)/T_b)}{\pi(t - T_b)/T_b}$$

Therefore

$$h(t)_{\text{Duobinary}} = \frac{T_b^2 \sin(\pi t/T_b)}{\pi t(T_b - t)}$$

The receiver for a duo-binary encoded signal simply involves the subtraction of the current decoded binary digit from the delayed previously decoded binary digit in an inverse process to the transmitter encoder. There is, however, a major problem with this technique if an error occurs in the decoding process. Because the output data bits are decoded using a previous data bit, if this is in error then the new output will be in error, and so on. In other words, errors will *propagate* through the system.

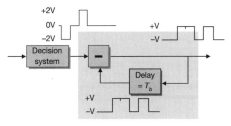

Duo-binary decoder

A clever technique for overcoming this error propagation problem is to use a *pre-coder* in the transmitter based on an exclusive-or gate as shown. The receiver then surprisingly becomes a simple rectifier followed by a threshold detector set at $+V$ volts with no error propagation problem. For more information on this technique see Carlson (1986).

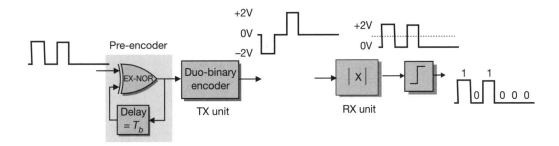

Modified duo-binary signalling

A useful extension of the duo-binary signalling method is *modified duo-binary* which has an overall frequency response which contains a spectral null at dc. This approach is thus well suited to channels which have poor low-frequency response (see Section 4.4, where an alternative, Manchester coding, is discussed).

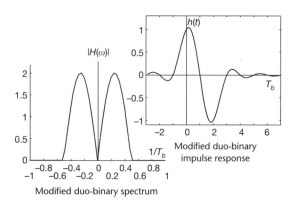

Modified duo-binary spectrum

The encoder involves a two-bit delay, causing the intersymbol interference to spread over two symbols. As for the basic duo-binary method, error propagation necessitates the use of a pre-coder which can again be implemented using an exclusive-or gate.

Detection of the modified duo-binary signal with pre-coding also involves rectification and a threshold set at $+V$ volts as before.

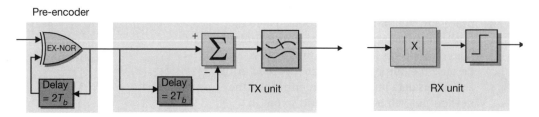

Modified duo-binary signalling system

IN DEPTH

Calculation of impulse response and frequency response for modified duo-binary signalling

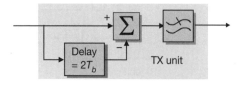

Modified duo-binary signalling system

The transfer function of an ideal delay element T_b is $e^{-j2\pi f T_b}$. The transfer function of the ideal low pass filter is simply:

$$H(f)_{LP} = \begin{cases} 1 & \rightarrow |f| \leqslant 0.5 \times 1/T_b \\ 0 & \rightarrow \text{otherwise} \end{cases}$$

Thus the overall frequency response of the modified duo-binary scheme is:

$$H(f)_{\text{Modified-Duobinary}} = H(f)_{LP}[1 - e^{-4j\pi f T_b}]$$
$$= H(f)_{LP}2j \cdot \sin(2\pi f T_b) e^{-2j\pi f T_b}$$

Therefore

$$H(f)_{\text{Modified-Duobinary}} = \begin{cases} 2j \cdot \sin(2\pi f T_b) e^{-2j\pi f T_b} & \rightarrow |f| \leqslant 0.5 \times 1/T_b \\ 0 \end{cases}$$

The impulse response $h(t)$ for the duo-binary scheme is simply the subtraction of two *sinc* *waveforms*, delayed by two bit periods with respect to each other:

$$h(t)_{\text{Duobinary}} = \frac{\sin(\pi t/T_b)}{\pi t/T_b} - \frac{\sin(\pi(t - 2T_b)/T_b)}{\pi(t - 2T_b)/T_b}$$

Therefore

$$h(t)_{\text{Duobinary}} = \frac{2T_b^2 \sin(\pi t/T_b)}{\pi t(2T_b - t)}$$

QUESTIONS

3.1 Which of the filter responses shown here exhibit the correct properties for achieving zero intersymbol interference in a filtered data scheme?

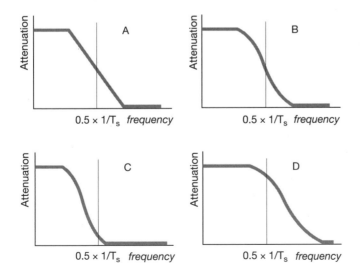

3.2 Sketch the eye diagram for a polar 4-ary Amplitude Shift Keying demodulated waveform, assuming noise-free operation with:

(a) an infinite operating bandwidth and no pulse shaping

(b) a root raised cosine channel with $\alpha = 0.5$.

3.3 A baseband binary data link is capable of supporting a bit rate of 4800 bps when using a raised cosine filter with an α of 0.6. How much faster could information be sent if the value of α was reduced to 0.2?

3.4 A 16 symbol state baseband modulation technique requires an α of 1 for reliable transmission. What is the maximum data rate that can be supported on this link, assuming a noise-free channel and a bandwidth of 3200 Hz?

3.5 A baseband cable modem is able to achieve a symbol error rate of 1 in 10^6 with binary signalling. With reference to the plot of symbol error rate vs E_b/N_0 for M-ary signalling (page 69), determine the approximate error rate that will result for the same E_b/N_0 value if a four-level modulation format were to be deployed.

3.6 A 64-level modulation format is measured as giving a symbol error probability of 2 in 10^5 at an E_b/N_0 value of 23 dB. What is the approximate bit error rate for the system, assuming that Gray coding has been used?

4 Sources and examples of channel degradation

4.1 Introduction

Having dealt in Chapter 3 with many of the issues surrounding practical filtering for digital communications systems in an idealized channel context, this chapter is aimed at injecting a bit of reality into the design process by considering the characteristics of a range of typical channels.

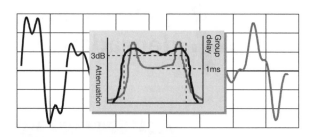

All communications links are ultimately limited by background noise in the system, and so it is essential to have a working knowledge of the *statistical properties* of noise as they affect data communications performance. Also, many channels are subject to interference, usually man-made, which can be equally detrimental to communications integrity. Thirdly, no communications link is truly distortion free, whether this be caused by imperfections in the processing hardware or defects within the channel, and the nature and impact of distortion on a digital communications system must be understood if good design choices are to be made.

In order to 'root' the discussion of noise, interference and distortion in reality, this chapter concludes with sections outlining the characteristics of two channel types in everyday use: the telephone channel, and the radio channel.

4.2 Gain, phase and group delay distortion

Gain distortion – filters

Most communications hardware – filters, mixers, amplifiers, and so on, and most channel types, cable, fibre, radio, infra-red, introduce amplitude distortion into a signal, usually with a frequency-dependent response.

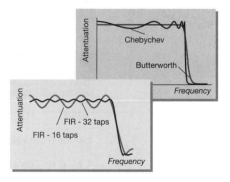

Filters, for example, are never perfectly 'flat' in the passband but all exhibit some degree of 'amplitude ripple'. Some filter types such as Elliptic or Chebychev filters have very high passband ripple, but also achieve very fast roll-off. Other filters such as Butterworth or Bessel filters have much less ripple but also much slower roll-off.

The raised cosine filter (see Section 3.4) also exhibits passband ripple, depending on the length (number of taps) used in the filter implementation. The degree of ripple can be made arbitrarily small with very long filter length, but at the expense of processing delay and complexity. In practice, the amplitude ripple in a digital raised cosine filter does not significantly degrade modem performance; however, other filtering within the communications system such as the crystal or ceramic filters used in radio IF (intermediate frequency) circuits can exhibit gross levels of amplitude distortion and this effect cannot be ignored.

Gain distortion – amplifiers

Amplifiers, particularly RF power amplifiers in radio systems and higher power lasers in optical systems, do not exhibit a linear relationship between input power and output power, often owing to *gain compression* in the devices used.

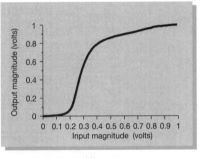

Typical amplifier gain response

The effect on digital signals passing through these non-linear devices is twofold. Firstly, the careful pulse shaping achieved with a Nyquist filter can be corrupted, reintroducing intersymbol interference (see Section 3.2) into the link. Secondly, the non-linearity can result in what is commonly termed *spectral regrowth* due to *intermodulation products* being generated within devices.

In many applications, particularly digital cellular systems, spectral regrowth is a major problem and a lot of design effort is focused on compensating for this problem, either by *linearizing* the components (see in-depth section), or by selecting modulation formats such as GMSK that are particularly tolerant to amplitude non-linearity.

In fact most components, mixers, op-amps, couplers and so on are not perfectly linear and hence introduce amplitude distortion; however, the distortion is usually much less than that introduced by the higher power or active RF components within a system.

IN DEPTH

Linear amplifiers/transmitters

At the heart of many of today's high performance digital wireless products is a transmitter whose linearity specification is becoming more and more exacting. Not only are these specifications calling for excellent linearity, but the close-in and wideband noise performance is also becoming more stringent and the anticipated/expected efficiency of transmitters is rising.

Techniques for linearizing amplifiers or transmitters fall largely into two categories – feedback or feedforward topologies. A short description of each of these techniques is given here, highlighting the applications of each.

Cartesian loop

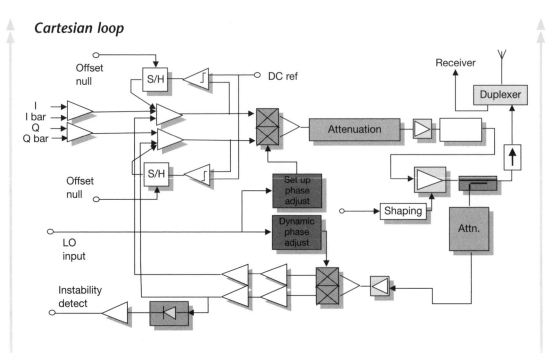

Cartesian loop

One of the most widely used linearization techniques in the market-place is based on Cartesian feedback which controls the linearity of an entire transmit chain, including the up-converter and amplifier stages. This technique is well suited to integration with a few Cartesian ICs already available in the market-place.

The nature of the feedback process means that the amount of correction afforded by the control system decreases with increasing modulation bandwidth. Typically, intermodulation improvement in the order of 30 dB is possible over a 25 kHz modulation bandwidth, and more than 10 dB improvement has been demonstrated for CDMA-type bandwidths in excess of 1 MHz. Solutions that meet the ETSI TETRA specifications have been designed. Efficiencies of Cartesian loop transmitters can be as high as 70%. Further information can be found in Wilkinson and Bateman (1989).

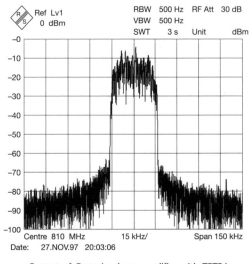

Ref Lv1	RBW	500 Hz	RF Att	30 dB
0 dBm	VBW	500 Hz		
	SWT	3 s	Unit	dBm

Centre 810 MHz   15 kHz/   Span 150 kHz
Date: 27.NOV.97 20:03:06

Output of Cartesian loop amplifier with TETRA
modulation input

RF synthesis/CALLUM

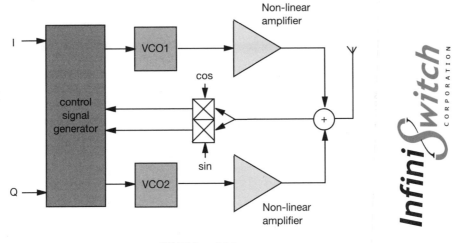

CALLUM modulator

The RF synthesis approach to transmitter linearization is a relatively new method which has the potential to achieve levels of efficiency approaching 80% without sacrificing intermodulation suppression or noise performance. CALLUM (Combined Analogue Locked Loop Universal Modulator) is a recent invention allowing closed loop control of the RF synthesis process. Simulation of this new technique suggests that this method can exceed the performance of Cartesian loop in all respects. Further information can be found in Bateman (1992).

Implementation of CALLUM transmitter

Pre-distortion

Fixed and controlled pre-distortion of a signal prior to amplification is a well-established technique for improving amplifier performance and recently a range of sophisticated adaptive pre-distortion designs have been published which aim to extend the capability of pre-distortion significantly.

Baseband adaptive pre-distortion techniques operate by attempting to generate a transfer function in the pre-distorter block that is complementary to the transfer function of the amplifier to be linearized, such that the combination has a linear input–output power characteristic. This approach is very similar to the channel equalization methods for overcoming gain and phase distortion on telephone channels or mobile radio channels. Further information can be found in Mansell and Bateman (1996).

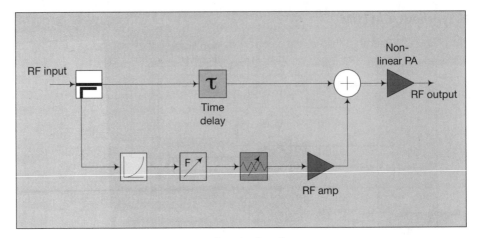

Pre-distortion lineariser

Feedforward

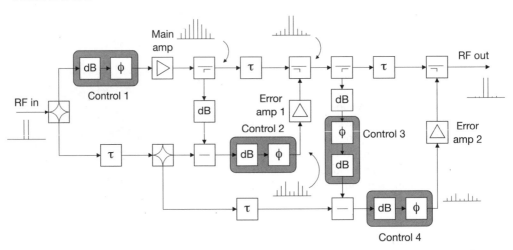

Dual loop feedforward amplifier

A rapidly expanding application of linear amplifiers is in multi-carrier applications, where flexibility of channel allocation, cost/performance benefit and modulation independence make this approach much more attractive than several single carrier amplifiers feeding a high power lossy tuned combiner network. The specification of linearity for GSM or PCS multi-carrier amplifiers is in excess of -75 dBc which requires very accurate intermodulation cancellation techniques.

Implementation of feedforward
amplifier

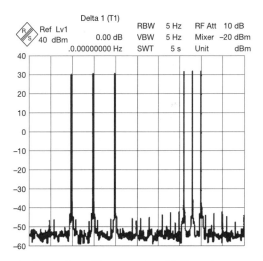

Feedforward amplifier output for −75 dBc performance

Single channel feedforward amplifiers are also of importance for wideband modulation formats such as IS-95 CDMA, where the challenge is to achieve linearity for minimal cost and high reliability. Further information can be found in Parsons and Kenington (1994).

Other techniques

Less well-known methods of amplifier and transmitter linearization include envelope elimination and restoration, polar loop correction, and LINK vector feedback. Further information can be found in Petrovic (1983).

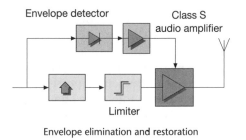

Envelope elimination and restoration

Gain distortion – the channel

The channel itself can introduce gain distortion via a variety of mechanisms. Wired channels exhibit a general roll-off of amplitude with frequency due to the capacitive nature of long lengths of cable causing a low pass filtering effect. Also, poor termination of cables at the end-user equipment can itself introduce distortion. Some telephone lines in particular exhibit a very poor amplitude response.

Wireless links do not suffer the capacitance problems of cables, but instead suffer from variable path loss due to the unpredictable nature of the communications path. This path loss can be either flat with frequency, or *frequency dependent*, depending on the bandwidth of the modulation signal, the operating frequency and the path distance. These issues are dealt with more fully in Section 4.5.

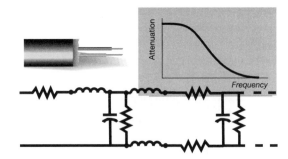

Phase distortion – filters

Just as filters introduce amplitude ripple in the passband, many also have phase variations across the passband and in the transition band. The effect of a non-flat phase response in a filter is to cause the various frequency components that make up the data signal to undergo slightly different amounts of *phase shift*. If the phase response is not flat, or does not increase linearly with frequency (termed a *linear phase response*), distortion will be introduced into the time waveform of the data pulse or symbol as shown here. If the component has a linear phase response,

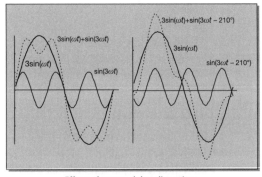

Effect of group delay distortion
(non-linear phase shift with frequency)

this translates into a fixed time delay of the signal passing through it, but otherwise the signal will *not* be distorted.

Again, some filters have better phase responses than others, with the *Bessel* filter having a very good near-linear phase response with frequency, and the *Elliptic* filter having a very poor response. *Digital* filters, on the other hand, can be implemented with *perfectly linear phase response* – so-called linear phase filters – and hence it is possible to implement a raised cosine filter (see Section 3.4) with no phase distortion using these techniques. (Note, not all digital filters have a linear phase response.)

Group delay – filters

'Group delay' is defined as 'the rate of change of phase shift with frequency'. For a filter with a linear phase response, the rate of change of phase with frequency is constant and hence the group delay will be a fixed, though non-zero, value over all frequencies, that is, a pure time delay.

For a non-linear phase response, the group delay will vary with frequency as shown here, with the effect that data pulses or symbols passing through the filter are smeared by the different delays across the frequency components, reintroducing ISI into the signal.

Both gain and phase/group delay distortion can be compensated to some extent with the use of equalization circuits, which in practice are themselves digital filters which are configured to compensate for the imperfections of the usually analogue filters within the system, or imperfections in the channel itself.

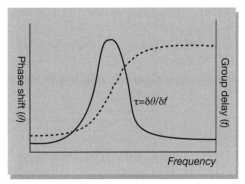

Origin of group delay

Phase distortion – amplifiers

Just as high power sources such as amplifiers and lasers have non-linear amplitude response with input power, they also usually have a non-linear phase response with input power. The phase response with frequency, however, is in practice quite linear and does not pose a problem unless very wide modulation bandwidths are being employed.

The effect of phase change with power level is to cause an amplitude-dependent phase distortion of the data signal, often termed *AM-PM distortion* (Amplitude Modulation to Phase Modulation). This has a particularly detrimental effect on phase-based modulation formats such as M-ary PSK (see Section 6.4) or M-ary QAM (see Section 6.5) which suffer unwanted rotation of the phase states of each symbol.

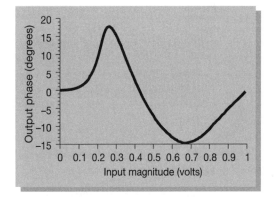

Phase distortion – the channel

The filtering effect of most wired channels will introduce some phase and hence group delay distortion into the signal, and here equalizers play the key role in measuring and compensating for much of this distortion.

In optical fibre links, distortion is usually attributed to time dispersion in the fibre, which gives rise to an upper limit on the data rate that can be supported. The time dispersion, which is caused by rays

Time dispersion in optical fibres

passing through the fibre with different numbers of reflections (called *multi-mode propagation*), is most effectively dealt with by using a single or *mono-mode fibre* and corresponding laser source, although these are more costly to manufacture.

In the wireless channel, phase distortion is not a problem for signals travelling along any individual path between source and receiver. In applications where the signal travels by multiple paths, however, each of which may have a different path length and hence relative time delay, the phases of the components making up the composite received signal will all be different. This can create significant problems for data signals as discussed in Section 4.5.

Frequency errors

Frequency errors within a communications link are caused by two mechanisms – inaccuracy of the frequency sources used in the modulation and demodulation process, and Doppler shift caused by the relative motion between source and receiver or from moving reflectors.

For modulation formats that use frequency or phase to represent different symbols, such as M-ary PSK (Section 6.4) or M-ary QAM (Section 6.5), frequency errors can lead to significant performance degradation for the modem.

Local oscillator error

For any bandpass modulation process it is necessary to generate sinewaves in both transmitter and receiver systems, ideally with precisely the same frequency and phase. This begs the question: 'Is it possible to implement two or more sinewave generators with perfect frequency and phase accuracy under realistic operating conditions (temperature variation, supply variation, ageing and so on)?'

A reasonably cost-effective crystal oscillator (<$10) may have a stability of 1 ppm (part per million) over a given temperature range. This means that for a telephone modem with a carrier of say 2 kHz, the oscillators at each end of the link could have an error of

$\pm 1.10^{-6} \times 2.10^{3} = 0.002\,\text{Hz}$. If we could ensure that both transmit and receiver carrier oscillators begin with the same phase, then we can expect the phase error between them to reach 360° after $1/0.004 = 250$ seconds, and 90°, giving zero output, after 75 seconds. These figures suggest that, providing an initial phase correction can be achieved, near phase-coherent detection can be ensured for a few seconds without further phase correction being required.

If we now consider the case of a cellular radio modem operating with a carrier of 1 GHz, then the oscillator frequency error for a 1 ppm source is $\pm 1000\,\text{Hz}$. Here, it is clear that simply achieving a correct starting phase will not allow us to ensure adequate coherency for more than a few microseconds. In this application, it is necessary to find a method of correcting the receiver carrier oscillator frequency and phase to match that of the transmitter. This process is termed *carrier recovery*.

Doppler shift

Whenever a signal source moves towards or away from a receiver, the frequency of the signal as observed at the receiver increases or decreases respectively. This is known as the *Doppler effect*. The degree of frequency shift is a linear function of the speed of motion and the carrier signal frequency. For example, a source moving at 70 mph using a carrier frequency of 900 MHz will experience a Doppler shift of up to $\pm 100\,\text{Hz}$ at the receiver.

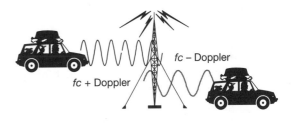

Correcting for Doppler shift can be very difficult, particularly in a multipath environment (see Section 4.5), where signals arriving from different angles experience different Doppler shifts.

IN DEPTH

Doppler shift

The Doppler shift introduced into a signal between transmitter and receiver units, or moving reflectors, is a function of their relative motion, the angle of arrival of the signal, and the operating frequency or wavelength. These parameters are related as follows:

Doppler shift (Hz) $= v \cdot f \cdot \cos(x)/c$

where v is the relative speed of TX and RX units in m/s, f is the carrier frequency in Hz, $c = 3 \times 10^{8}$ m/s and x is the relative angle of arrival in degrees.

For example, the Doppler shift experienced by a cellular phone within a car travelling at 70 mph directly away from a base-station transmitter, operating on the DCS1800 system (operating frequency 1.8 GHz) is 189 Hz. The Doppler shift for a person walking along the street at 4 mph is, however, much smaller at only 11 Hz.

A typical spectrum of a signal received at a mobile terminal when moving within a multipath environment with several reflectors each giving different angles of arrival and hence different Doppler shift is shown in the figure. The source was a single tone at the carrier frequency.

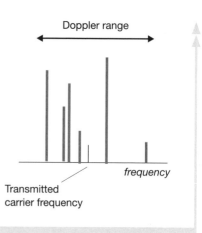

4.3 Interference and noise

Sources of interference

Most interference encountered in digital communications systems (except for deep space missions!) arises from either other communications systems or machinery. For example, crosstalk in telecommunications lines is classed as interference, as is the 'ignition noise' generated by a car engine.

In radio systems in particular, a major source of interference is from other users of the radio spectrum. For example, equipment radiating on frequencies close to the wanted channel can pass through the receiver selection filtering, causing what is termed *adjacent channel* interference. In cellular applications, mobile users in different geographical locations are assigned the same frequency for their calls, and if they are not separated by sufficient distance, *co-channel interference* occurs.

In both radio and television, multipath interference is common, manifest as ghosting on the television screen, caused by signals travelling by many different paths between transmitter and receiver each with slightly different time delay.

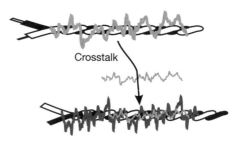

Crosstalk

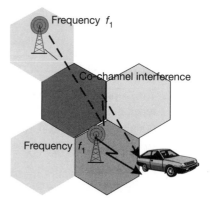

Frequency f_1

Co-channel interference

Frequency f_1

Dealing with interference

Because most interference (excluding noise) in communications systems is generated by other items of equipment, it is often possible with good design to minimize the effects of interference. This can be achieved both by judicious selection of modulation and coding format to be the least sensitive to a given type of interference, and also by tackling the causes of the interference.

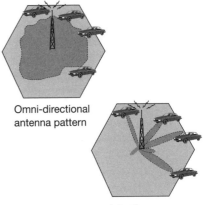

Omni-directional antenna pattern

Adaptive array antenna pattern

Crosstalk in telephone lines, for example, can be reduced by careful layout of cables, or by replacing cables with optical fibre, which has no external radiation to cause crosstalk.

Ghosting caused by multipath can often be cured by using directive antennas to avoid picking up reflections. Some modern cellular base-stations and even some mobile handsets are using adaptive antennas which in real time change the direction of the antenna beam to 'null out' interferers and 'focus' on the wanted signal.

Co-channel and adjacent channel interference is again controllable by good system planning and good selective filtering within the receiver modem.

Sources of noise

Unlike interference, noise originates predominantly from within the communications link itself and is usually totally random in nature, making it very difficult to deal with. There is a variety of mechanisms by which noise is generated, the most commonly referenced forms being thermal noise, shot noise, flicker noise and atmospheric noise.

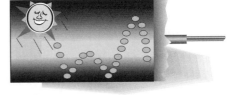

Thermal noise often dominates in communications systems and originates from the free movement of electrons within a conductor. The name arises because the energy and hence degree of movement of electrons increases proportionally with the temperature of the conductor. The current and hence voltage generated by this movement has a waveform that is entirely random in nature and which will, over time, have an average power spectrum that is flat over all frequencies. This property of thermal noise to contain all frequencies has resulted in it being called 'white noise' to mirror the property of white light to contain all colours.

A good text on the subject of noise in digital communications systems is by Schwartz (1990).

Thermal noise

The *average power attributable to thermal noise* is:

Thermal noise $N_{av} = kTB$

where k is Boltzmann's constant $= 1.38 \times 10^{-23}$ Watts/Hz/°K, T is the absolute temperature in degrees Kelvin, and B is the bandwidth in which the measurement is made.

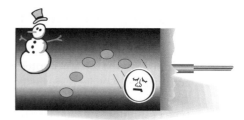

Thermal noise can clearly be reduced by cooling the noise source and this very principle is being applied in some radio receivers using cryogenic coolers, to improve the receiver sensitivity.

EXAMPLE 4.1

A radio receiver is limited in performance by thermal noise in the receiver 'front-end'. It is designed to provide an S/N ratio of better than 10 dB to the demodulator input. The channel bandwidth for the receiver is 25 kHz. What is the minimum received signal level that can be used to achieve this performance target, assuming the receiver is operating at a temperature of 280°K?

Solution

The average power of thermal noise for this case is given by:

$$N_{av} = KTB = 1.38 \times 10^{-23} \times 280 \times 25\,000 = 9.66 \times 10^{-17}\,\text{W}$$

or -130 dBm (dB relative to 1 mW).

To achieve an S/N ratio of 10 dB, the received signal power must therefore be in excess of $(-130 + 10) = -120$ dBm.

Shot, flicker and atmospheric noise

Shot noise is generated within semiconductor junctions when electrons cross a potential barrier. Whereas thermal noise power is proportional to temperature, shot noise power is proportional to the bias current in the semiconductor. The nature of shot noise is also purely random and has a flat power spectrum with frequency.

Flicker noise is also generated in semiconductors and is proportional to the dc bias current, but differs in that the noise power decreases with frequency. Because this power variation is almost directly proportional to $1/f$, it is sometimes called *1/f noise*.

Atmospheric noise is a general term given to noise arising from electromagnetic radiation from solar and galactic sources. Certain stars, for example, emit definite and regular amounts of noise which are best avoided by pointing the antenna away from the noise source. The compound effect of this noise is usually expressed as an equivalent *sky noise temperature* and is generally much less than thermal noise. The level of noise varies considerably with frequency, with the higher levels of noise occurring in the microwave region of the spectrum.

Characteristics of noise

Noise is usually classified as *white* or *coloured* depending on the spectral density of the noise power with frequency.

White noise is defined as having a flat power spectral density over all frequencies of interest, with a value usually denoted as N_0 Watts/Hz.

Coloured noise has a non-uniform spectral distribution; however, over a finite bandwidth corresponding perhaps to a single communications channel, the power spectral density might appear flat, hence the term *bandlimited white Gaussian noise*.

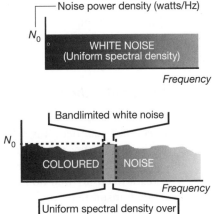

Not only is it necessary for the spectrum to be flat, but the statistics of the noise must be such that the envelope distribution of the bandlimited noise must be Gaussian in nature to fully satisfy the Shannon condition (see Section 2.4). Fortunately this holds approximately true for the majority of practical narrowband communications systems.

4.4 The telephone channel

Ac-coupled channels

In the discussions so far, we have assumed that the channel is able to pass all frequency components of a data signal from 0 Hz up to a bandwidth of B Hz.

There are, however, many communications channels that are not able to pass low frequency components, either because they need to be *ac-coupled* for practical reasons (such as eliminating unwanted dc-offsets), or because the bandwidth allocated to a given user is in a part of the spectrum well removed from 0 Hz, for example wireless communications.

The domestic telephone channel is a classic example of a channel which is *bandpass* in nature, that is, it has a low frequency and high frequency cut-off in its gain response. The low frequency cut-off (ac-coupling) results from capacitive and/or inductive coupling of the telephone line at both the exchange and subscriber ends allowing dc power for the telephone to be passed over the same cable as the speech or data signal. The high frequency cut-off is a combination of deliberate filtering at the exchange to minimize noise on the channel and also the transmission line filtering effect of long lengths of cable.

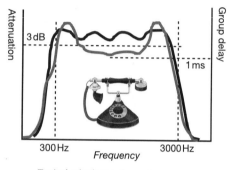

Typical telephone channel response

A typical telephone channel response is shown here. Notice that even within the channel, the gain response may not be flat and this in itself will introduce further symbol degradation.

NRZ vs Manchester encoding

In order to allow unmodulated data symbols to pass uncorrupted through ac-coupled channels, it is necessary to ensure that the symbol stream has very little or no energy at or near 0 Hz. This means that the data must be either *scrambled* or *encoded* to ensure that no long strings of 1s or 0s are transmitted.

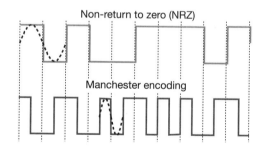

Scrambling of the data involves reordering of the data symbols in such a way that the chances of long strings of 1s or 0s occurring in the scrambled waveform is minimized or eliminated. This process introduces a processing overhead and also a latency in the encoding and decoding process which may introduce unacceptable time delay in the communication process – particularly if digital voice is being carried. In most cases, a data encoding process is used in preference to scrambling, albeit at the expense of occupied bandwidth in most cases.

There are many data encoding schemes for reducing the dc content of a data signal, the most common of which is *Manchester encoding*. It can be seen that compared with the original *non-return-to-zero* (NRZ) data stream, the Manchester encoded signal never contains long strings of 1s or 0s.

For a more detailed discussion of line coding techniques, see Haykin (1989).

Bandwidth efficiency of encoded data

The spectrum of a Manchester encoded signal clearly demonstrates the effectiveness of the coding process in eliminating the dc content of the signal, but also shows an approximate *doubling of the bandwidth* compared with the NRZ data stream as would be expected from the increased rate of change of waveform.

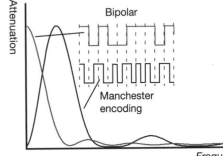

The bandwidth efficiency of Manchester encoded binary data thus drops from 2 bits/second/Hz for the NRZ signal to only 1 bit/second/Hz.

> Bandwidth efficiency of Manchester encoded binary data = 1 bit/second/Hz

E EXAMPLE 4.2

A cabled binary baseband data system that supports a maximum information transfer rate of 1200 bps using bipolar signalling has to be modified to allow ac coupling of the receiver unit owing to dc offset problems in the circuit design. How might the system be altered to fulfil this task, and what would be the impact on the information transfer rate?

Solution

An effective ac coupling method is Manchester encoding of the data which eliminates long strings of 1s or 0s. Manchester encoding, however, doubles the occupied bandwidth of the data signal and hence the usable data rate would drop from 1200 bps for bipolar signalling to 600 bps for Manchester encoded data transfer.

Data encoding and the telephone channel

While data encoding can effectively move the signal energy away from 0 Hz and permit ac-coupled operation, the encoding process does not provide sufficient shifting of the frequency components within a baseband data symbol stream to allow them to pass effectively through the bandpass response of the telephone link.

In order to ensure minimum distortion of the data passing over the telephone channel, it is clear that the spectral energy of the data signal must be placed in the centre of the channel passband where there is the flattest gain response and the smoothest group delay. This requirement is achieved by modulating the baseband data stream onto a carrier signal and is described in Chapter 5.

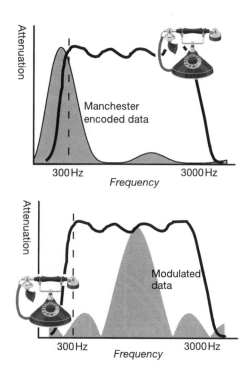

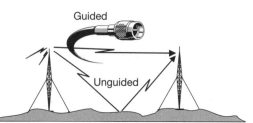

4.5 The wireless channel

Unguided propagation

The wireless channel is unique in that it is an *unguided medium* (unlike cable or fibre), and it is the means by which signals propagate from transmitter to receiver that dominates the data communications performance on a wireless link.

In essence there is little inherent filtering or distortion in a wireless link (other than frequency-dependent absorption in the atmosphere), provided that there is only one propagation path between TX and RX units. This can only be achieved by ensuring that there are *no reflections* of the transmitted signal arriving within the 'aperture' of the receiver. One of the few examples where this holds true is for satellite to ground communications where the receiving antenna is a very focused parabolic reflector pointing directly towards the satellite and there are no objects (other than distant planets!) from which a reflection can arise.

Multipath distortion

In applications where more than one propagation path will exist, the interaction of the signals from these multiple paths at the receiver (multipath propagation) gives rise to significant distortion of the received data symbols.

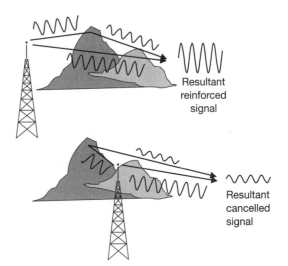

Resultant reinforced signal

Resultant cancelled signal

The same source signal, arriving by a different route, will experience a different path length and hence a different propagation delay. This difference in delay will result in different phases between the two received signals. If the phase difference approaches 180° then the signals will in fact partially cancel each other, while if the phase difference approaches 0° they will reinforce.

Multipath fading

If either the transmitter, receiver or reflectors are moving within a multipath environment, the path lengths will vary with time and so the relative phases between signals will also vary with the position of the users. The result is that the receiver experiences a combined signal with fluctuating amplitude and phase as a function of time.

The depth of the resulting fading of the signal is very dependent on whether a strong 'line-of-sight' path exists between transmitter and receiver, in which case the fading signal envelope usually conforms to a *Ricean statistical distribution*, whereas if the line-of-sight path is obscured, and energy arrives at the receiver by a large number of reflections, then a *Rayleigh distribution* of the envelope level is more likely to occur. The characteristics of the fading signal envelope make a significant impact on the bit error rate performance for a digital communications link. An excellent book on multipath fading is Jakes (1993), and a very thorough text on the performance of data modems in fading is given in Proakis (1989).

Frequency flat vs frequency selective fading

A challenging phenomenon of multipath propagation is that the degree of cancellation or reinforcement of signals changes for the different frequencies within a data signal. This is because the relative phase shift between two frequency components undergoing identical path delays will be different as the wavelength of the two components is altered. The effect is known as *frequency selective fading* and gives rise to notches in the frequency response of the channel.

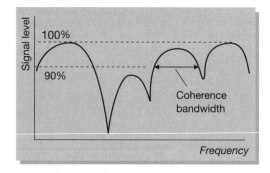

If the path lengths for all the signals are very similar compared to the wavelengths of the signal components, then the phase variations between components will be small and they will all undergo very similar amounts of reinforcement or cancellation. This is often termed *flat fading*. On the other hand, if the difference in path length is large, the fading characteristics will vary greatly even with small frequency separations.

The degree of correlation between fading over a range of frequencies is expressed in terms of the *coherence bandwidth* which is defined as the bandwidth over which the fading statistics are correlated to better than 90%.

Multipath fading – the time domain problem

If the multipath delays, typically a few microseconds for outdoor communications, are significant with respect to the symbol period, then *intersymbol interference* will occur. Note: this effect of multipath propagation is not an additional form of distortion to frequency selective fading but rather the same effect viewed in the time domain.

To overcome the problem, channel equalizers are often employed which need to respond (adapt) to the changing nature of the channel. Alternatively, the symbol rate must be reduced by using M-ary signalling or many parallel channels so that the delay becomes a less significant proportion of the symbol period (typically <10%).

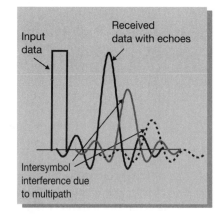

It should be noted that propagation delay itself is not a major problem, but rather the *spread of delays* over the different propagation paths. Shown here is a typical plot of the amplitude and delay profile for a cellular telephone signal as it propagates within a city centre. The spread of time delays is about $15\,\mu s$ for the significant components.

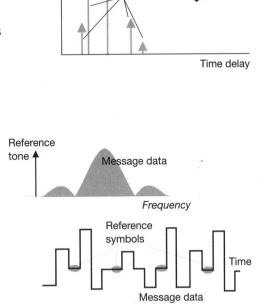

Coping with multipath fading

Reference sounding

Multipath fading is both a blessing and a curse for wireless digital communications systems. Were it not for the echoes of the transmitted signal caused by reflections from buildings, signals would not reach the user nestled within the crowded shopping street where the line-of-sight path to the transmitter is totally obscured. On the other hand, mitigating against distortion caused by multiple signals arriving at a receiver is a non-trivial task.

If the fading is 'flat' with frequency, then a common method of combating the multipath-induced amplitude and phase variations is to use a 'reference sounding signal'. By sending a known frequency tone or periodic known data symbols alongside the message data, these references can be used to measure in real time the instantaneous amplitude and phase variations imposed by the channel (Bateman, 1990). For flat fading, it can be inferred that similar gain and phase distortion are imposed on all other frequency components in the message signal. The information gleaned from the reference can thus be used to subtract out the distortion from the message data. It is generally assumed that the multipath fading will be 'flat' with frequency over a bandwidth of less than 25 kHz for most mobile radio applications.

IN DEPTH

Pilot tone-based fading correction

A number of mobile radio systems use either pilot tone(s) or pilot symbols for correction of narrowband multipath fading. When correctly applied, the information provided by the reference tone concerning the instantaneous Doppler shift and amplitude/phase fluctuations

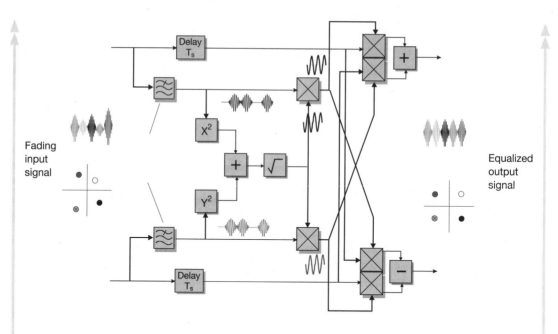

Feedforward signal regeneration

of the received signal due to multipath can be used to fully correct for distortion in the wanted data signal.

Because the fading pattern can change several hundred times per second for some mobile data applications, the pilot-based correction processing must operate in real time. This requires feedforward correction techniques, of which a commonly used method is feedforward signal regeneration (FFSR). A circuit diagram for a quadrature implementation of an FFSR circuit ideally suited for M-QAM modulation is given here. Further information can be found in Bateman and McGeehan (1983).

Parallel transmission

Where the time delay spread is such that the fading is not flat with frequency, reference sounding techniques will not work unless several *parallel data channels* are used, each of which occupies a narrow bandwidth over which the fading is 'flat' with each channel having a separate sounding reference.

If the concept of parallel data channels is taken to an extreme, and coding redundancy

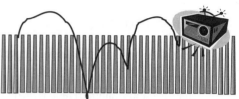

Frequency

is added to the data source such that the data integrity is not compromised when some of the parallel channels fall within the fading notches (remember these notches change position with time as the user moves), then it is possible to do away with the reference signals and simply rely on sufficient of the parallel channels being decoded correctly to allow the receiver to correct for any data errors in the few faded sub-channels.

This very technique is being adopted for digital audio broadcast (DAB) and digital video broadcast (DVB) in Europe where up to 1024 parallel data channels are used to convey the digitized music, voice and images to the radio or television screen. The modulation format is termed Orthogonal Frequency Division Multiplexing (OFDM).

Spectral spreading

Two alternative methods exist for coping with both flat and frequency selective fading; both involve increasing the bandwidth required to send the data.

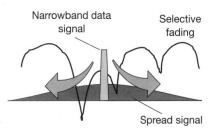

A technique known as direct sequence spread spectrum (see Section 8.4) uses a wideband data sequence to mix with a narrowband data signal and thence *spread* the energy well outside the coherence bandwidth for the channel. A small proportion of the spread signal energy will be lost in the frequency selective fades, but the majority will pass with little distortion through the channel. By *de-spreading* the signal in the receiver, a reasonable copy of the original transmitted signal can be obtained. Data coding and channel equalization are often employed in addition to the spreading to improve the integrity of the channel.

Instead of spreading the data signal instantaneously (and thinly) over a wide frequency range, an equally effective method is to rapidly change the position of the narrow data signal within a much wider bandwidth. This frequency hopping (see Section 8.4) approach means that for some of the time the signal will fall within a selective fade, but for most of the time, it will be passed within a non-fading portion of the channel. The result, as for direct sequence spreading, is that most of the data signal most of the time reaches the receiver with little distortion, and with extra coding a high integrity communications link can be established.

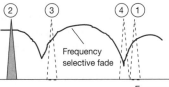

Channel equalizers

For high-speed wireless data applications where pilot sounding or parallel transmission is not appropriate or there is not sufficient bandwidth available to spread the signal effectively, then it is necessary to use wideband channel equalization techniques. These in effect involve sending a sounding pulse into the channel and measuring the level, phase and time delay of each significant echo received from the various transmission paths. The receiver then has to work out the 'inverse channel transfer function' with which to correct the subsequent message data.

Because the strength, number and delay of the echoes vary with time as the user moves, this channel sounding must be repeated frequently and a new inverse channel transfer function calculated each time. A modern example of this approach to combating multipath is contained

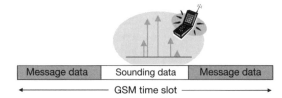

| Message data | Sounding data | Message data |

GSM time slot

within the GSM digital cellular system. The modulation bandwidth for GSM is 200 kHz which means that the fading is frequency selective and equalization is necessary. A sounding data sequence is sent embedded in the centre of each GSM data packet (frame) which is repeated every 4.615 ms. In other words, the cellular phone is working out the echo response of its surroundings at a rate of approximately 200 times per second.

Directional antenna

One logical way to reduce the effect of multipath fading is to reduce the number of paths seen by the receiver. This in practice means using a *directional antenna* on either the transmitting unit or receiving unit, or both.

Unfortunately, it is very difficult to make an antenna directive at low radio frequencies (below a few hundred MHz), without requiring a very large physical structure. As the frequency of operation increases, the physical size of the antenna can be reduced in line with the reducing wavelength, and directivity is much more easily realized.

For example, television aerials achieve modest directivity by having several elements in an array and this leads to a significant reduction in 'ghosting' (caused by multipath echoes) of the signal on the television screen. Microwave links achieve very good directivity by using a parabolic reflector.

It is clearly not sensible to put a fixed directional antenna on a moving object, the orientation of which will change with time, and under these circumstances, the only solution is to try to implement an adaptive directional antenna that can track the wanted signal over time.

Predicting multipath distortion

In order to design wireless digital communications systems that can best cope with the multipath environment in which they must operate, it is very helpful to be able to predict and simulate these effects in the system design phase. Apart from statistical prediction tools based on extrapolations from measured propagation data, one of the most accurate and versatile modelling techniques is based on *ray tracing*.

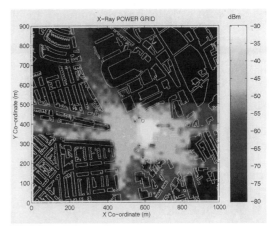

In a ray-tracing system, the *reflective*, *refractive* and *dispersive* properties of physical objects are modelled within a computer, along with their three-dimensional position. Every possible path that could be taken by a signal passing from transmitter to receiver is then traced by the computer, building up a picture of the level, angle of arrival and time delay of each one. From this information, it is possible to determine the frequency response of the channel, and to simulate the effects of the multipath fading on actual modem performance. This technique is very similar to that used to produce real-world lighting effects on virtual 3-D models created in a computer.

How propagation changes with operating frequency

The way in which signals propagate though the air varies with the frequency or wavelength. Some microwave frequencies, for example, are heavily attenuated by water droplets or oxygen molecules in the atmosphere when the size of these particles becomes comparable with the signal wavelength (60 GHz for water molecule absorption).

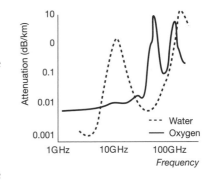

In contrast, at very low operating frequencies (<30 MHz) where the wavelength is very large, radio signals can propagate extremely long distances (in fact around the earth) owing to the various layers within the ionosphere acting as giant wave guides reflecting the signal to and fro between the layers. Unfortunately the properties of the ionospheric layers change with time of day, season, temperature and so on, and so the propagation characteristics can be very unpredictable and data transmission rates are often limited to a few kbps.

Above 30 MHz, ionospheric reflection begins to pack up and propagation is principally by line-of-sight path. In order to communicate over long distances, high antenna towers are required to combat the curvature of the earth.

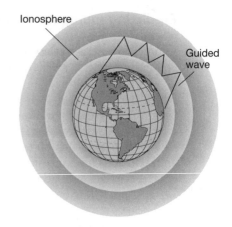

QUESTIONS

4.1 A cable is measured to have a flat gain response with frequency over the band of interest, but is found to have a phase response that changes proportionally with frequency, with a measured phase increase of 5° for every 1 MHz of bandwidth. What is the group delay response for the cable?

4.2 What is the average thermal noise power at a temperature of 17°C measured in a bandwidth of 20 kHz?

4.3 A seismic data telemetry link is required to operate over a temperature range of −40°C to +80°C. If the average thermal noise power at room temperature 17°C is found to be −126 dBm, how will this change at the extremes of the operating temperature range?

5 Bandpass digital modulation

5.1 Introduction

The previous chapters have been largely concerned with so-called *baseband signalling* where the channel band is assumed to extend from 0 Hz upwards. In applications where contiguous bandwidth encompassing 0 Hz is not available, *bandpass signalling* is required. Here, the task is to centre the symbol energy at a given frequency of operation, for example 900 MHz for a typical cellular telephone channel and 30 000 GHz (1000 nm) for an optical fibre link.

This process usually involves *modulating* the amplitude, frequency and/or phase of a *carrier* sinewave. The carrier is commonly written as $\cos(\omega_c t)$.

We shall see that the choice of modulation method crucially affects the ease of implementation, the noise tolerance and occupied channel bandwidth of the resulting bandpass data modem.

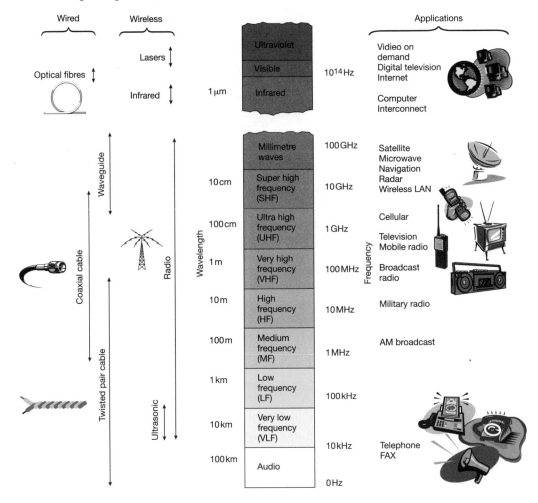

On the previous page, a chart is provided showing how the bandpass spectrum is allocated to different communications services based on the carrier frequency.

5.2 Amplitude Shift Keying (ASK)

What is ASK?

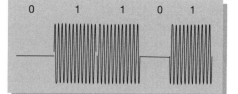

The simplest form of bandpass data modulation is *Amplitude Shift Keying* (*ASK*). Here, the symbols are represented as various discrete amplitudes of a fixed frequency carrier oscillator.

In binary ASK, where only two symbol states are needed, the carrier is simply turned on or off, and the process is sometimes referred to as *ON-OFF Keying* (*OOK*).

If more than two symbol states are used, then an M-ary ASK process is adopted, an example being the 8-ASK format shown here.

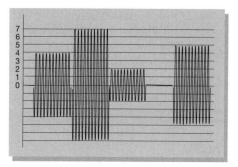

Symmetry in ASK

The spectrum of an ASK signal can easily be determined if the spectrum of the baseband data symbol stream is known, by viewing the ASK modulation process as a mixing or multiplication of the baseband symbol stream with the carrier component.

If we consider for the moment a single frequency component, $\cos \omega_m t$, from within the baseband spectrum, and perform the mathematical multiplication with the carrier, $\cos \omega_c t$, the modulated signal becomes:

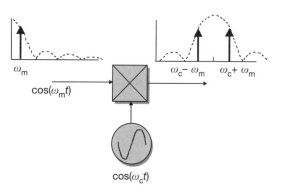

$$\cos \omega_m t \cdot \cos \omega_c t = 0.5 \cos(\omega_c - \omega_m)t + 0.5 \cos(\omega_c + \omega_m)t$$

The modulated spectrum for this sample component becomes two identical components spaced *symmetrically* either side of the carrier frequency.

Spectral occupancy of ASK: ASK data spectrum

If we now include all the components in the baseband stream which will mix with the carrier to generate a sum and difference component, the resulting spectrum is again symmetrical about the carrier frequency, and is in fact a positive and reversed image of the baseband 'sinc' spectrum for an unfiltered binary data stream.

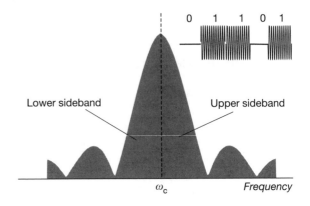

This ASK spectrum is sometimes referred to as a *double sideband spectrum*, with an *upper* and *lower* sideband with respect to the carrier. It is immediately evident that the bandwidth occupied by the ASK modulation is *twice* that occupied by the source baseband stream with a maximum bandwidth efficiency of:

> *Bandwidth efficiency of binary ASK* = 1 *bit/second/Hz*

EXAMPLE 5.1

An Amplitude Shift Keying format is used for transmitting data at a rate of 28.8 kbps over a telephone channel with bandwidth extending from 300 Hz to 3400 Hz.

(a) How many symbol states are required in order to achieve this level of performance?

(b) What would be the equivalent number of symbol states needed if the channel passband extended from 0 Hz to 3100 Hz and baseband M-ary signalling was used?

(c) What is the theoretical capacity for the ASK system if the S/N ratio on the telephone link is 33 dB?

Solution

(a) The capacity of a bandpass ASK channel is given by:

$$C_{ASK} = B\log_2 M$$

compared with

$$C_{baseband} = 2B\log_2 M$$

Hence,

$$28\,800 = (3400 - 300)\log_2 M$$

and

$$M = 626.1$$

or 1024 states to nearest power of 2.

(b) For the baseband equivalent,

$$28\,800 = 2 \times 3100\log_2 M$$

Thus

$$M = 25.02$$

or 32 states to the nearest power of 2.

(c) Applying the Shannon–Hartley equation,

$$C = B\log_2(S/N + 1)$$

we obtain:

$$C = (3400 - 300)\log_2(10^{3.3} + 1) = 33.996\,\text{kbps}$$

Note: the Shannon–Hartley expression is valid for both baseband and bandpass channels – it is the number of symbol states that must be increased on a bandpass channel, but as we will see for QPSK, this does not necessarily imply that the E_b/N_0 performance will be degraded compared to a baseband link.

Generation of ASK modulated signals

We have already seen that an ASK signal can be realized using a mixer to multiply the carrier with the baseband symbol stream. This is termed a *linear modulation* process.

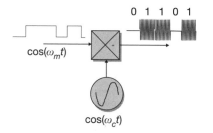

A simpler alternative, particularly for binary ASK, is to use a switch to gate the carrier on and off, driven by the data signal. For more than two symbol states, this approach becomes quite complicated with the requirement to gate on carriers with differing amplitudes to represent the required number of symbol states.

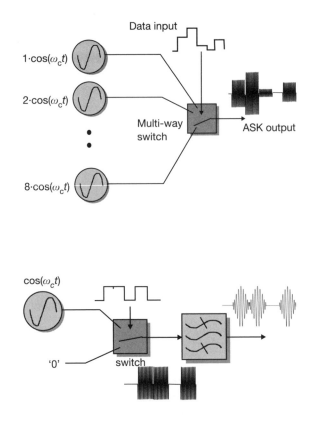

Bandwidth limited ASK

Bandpass filtering method

In order to minimize the occupied bandwidth of the transmitted ASK signal, filtering or pulse shaping (see Chapter 3) is required either prior to or after modulation onto a carrier.

The switching method of ASK generation does not allow any pre-filtering of the modulating baseband symbol stream, as the switch is a *non-linear process* and does not transfer the pulse-shaping information onto the carrier envelope. In this case, any filtering to constrain the bandwidth must be performed on the modulated bandpass signal.

For example, if we wish to filter a 30 kHz wide modulated signal super-imposed on a carrier at 900 MHz (typical of some digital cellular systems in the USA), this would imply a filter with a *Quality factor* $Q = 900 \times 10^6 / 30 \times 10^3 = 30\,000$. At present, this Q can only be achieved using crystal filters, which have very poor amplitude ripple and group delay distortion in the passband. They certainly do not allow the designer to achieve the root raised cosine bandpass (see Section 3.4) response required for zero intersymbol interference.

Baseband filtering method

The problems with bandpass filtering a high frequency modulated data signal can be eliminated if the pulse shaping is performed on the baseband input data stream, and a linear modulation process employed which preserves the amplitude information of the data signal.

Using the mixer-based approach the baseband data stream can be pre-filtered using a low pass (root raised cosine (see Section 3.4)) filter and this pulse-shaping information will be imposed (assuming the mixer is sufficiently linear) as the envelope variation of the carrier.

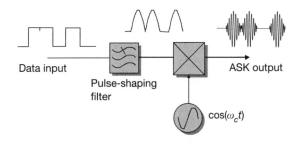

In practical terms, linear active integrated mixers capable of operating at carrier frequencies up to several GHz are now readily available off the shelf, costing only a few dollars (see in-depth section on pages 110–111).

Non-coherent detection

With ASK, the information is conveyed in the amplitude or envelope of the modulated carrier signal and the data can thus be recovered using an envelope detector. The simplest implementation of an envelope detector comprises a diode rectifier and smoothing filter and is classed as a *non-coherent detector*. The simplicity of this approach is offset by its reduced ability to differentiate the wanted signal from noise when compared with a coherent detector (see below).

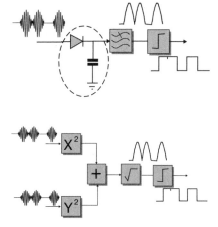

If quadrature versions of the modulated carrier signal are available in the receiver, that is, $a(t) \cdot \cos \omega_c t$ and $a(t) \cdot \sin \omega_c t$ (where $a(t)$ represents the data imposed amplitude modulation), then an alternative form of envelope detector can be used based on squaring and adding the two quadrature signals and then taking the square root. Mathematically we get:

$$a(t)^2 \cos^2 \omega_c t + a(t)^2 \sin^2 \omega_c t = a(t)^2(\cos^2 \omega_c t + \sin^2 \omega_c t) = a(t)^2$$

Coherent detection

A *coherent detector* operates by mixing the incoming data signal with a locally generated carrier reference and selecting the difference component from the mixer output. Representing the modulated data signal as $a(t) \cdot \cos \omega_c t$ and the reference carrier as $\cos(\omega_c t + \theta)$, the mixer output becomes:

$$a(t) \cdot \cos \omega_c t \cdot \cos(\omega_c t + \theta) = 0.5 \cdot a(t) \cos(\theta) + 0.5 \cdot a(t) \cos(2\omega_c t + \theta)$$

IN DEPTH

Example of integrated 'linear' mixer (in this case optimized for demodulation). Data sheet courtesy of Motorola.

MOTOROLA
SEMICONDUCTOR TECHNICAL DATA

Order this document
by MRFIC2001/D

The MRFIC Line
900 MHz Downconverter (LNA/Mixer)

MRFIC2001

900 MHz
DOWNCONVERTER
LNA/MIXER
SILICON MONOLITHIC
INTEGRATED CIRCUIT

The MRFIC2001 is an integrated downconverter designed for receivers operating in the 800 MHz to 1.0 GHz frequency range. The design utilizes Motorola's advanced MOSAIC 3 silicon bipolar RF process to yield superior performance in a cost effective monolithic device. Applications for the MRFIC2001 include CT-1 and CT-2 cordless telephones, remote controls, video and audio short range links, low cost cellular radios, and ISM band receivers. A power down control is provided to minimize current drain with minimum recovery/turn-on time.

- Conversion Gain = 23 dB (Typ)
- Supply Current = 4.7 mA (Typ)
- Power Down Supply Current = 2.0 μA (Max)
- Low LO Drive = −10 dBm (Typ)
- LO Impedance Insensitive to Power Down
- No Image Filtering Required
- No Matching Required for RF IN Port
- All Ports are Single Ended
- Order MRFIC2001R2 for Tape and Reel.
  R2 suffix = 2,500 Units per 12 mm, 13 inch Reel.
- Device Marking = M2001

CASE 751–05
(SO–8)

ABSOLUTE MAXIMUM RATINGS (T_A = 25°C unless otherwise noted)

Rating	Symbol	Value	Unit
Supply Voltage	V_{CC}	5.5	Vdc
Control Voltage	ENABLE	5.0	Vdc
Input Power, RF and LO Ports	P_{RF}, P_{LO}	+10	dBm
Operating Ambient Temperature	T_A	−35 to + 85	°C
Storage Temperature	T_{stg}	−65 to +150	°C

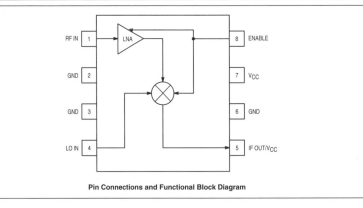

Pin Connections and Functional Block Diagram

REV 2

© Motorola, Inc. 1994

 **MOTOROLA**

RECOMMENDED OPERATING RANGES

Parameter	Symbol	Value	Unit
Supply Voltage Range	V_{CC}	2.7 to 5.0	Vdc
Control Voltage Range	ENABLE	0 to 5.0	Vdc
RF Port Frequency Range	f_{RF}	500 to 1000	MHz
IF Port Frequency Range	f_{IF}	0 (dc) to 250	MHz

ELECTRICAL CHARACTERISTICS (V_{CC}, ENABLE = 3.0 V, T_A = 25°C, RF @ 900 MHz, LO @ 1.0 GHz, P_{LO} = –7.0 dBm, IF @ 100 MHz unless otherwise noted)

Characteristic (1)	Min	Typ	Max	Unit
Supply Current: On-Mode	—	4.7	5.5	mA
Supply Current: Off-Mode (ENABLE < 1.0 Volts)	—	0.1	2.0	µA
ENABLE Response Time	—	1.0	—	µs
Conversion Gain	20	23	26	dB
Input Return Loss (RF IN Port)	—	13	—	dB
Single Sideband Noise Figure	—	5.5	—	dB
Input 3rd Order Intercept Point	– 26	– 22.5	—	dBm
Output Power at 1.0 dB Gain Compression	—	–10	—	dBm
LO – RF Isolation (1.0 GHz)	—	37	—	dB
LO – IF Isolation (1.0 GHz)	—	33	—	dB
RF – IF Isolation (900 MHz)	—	4.0	—	dB
RF – LO Isolation (900 MHz)	—	19	—	dB

NOTE:
1. All Electrical Characteristics measured in test circuit schematic shown in Figure 1 below:

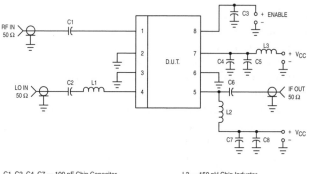

C1, C2, C4, C7 — 100 pF Chip Capacitor
C3, C5, C8 — 1000 pF Chip Capacitor
C6 — 6.8 pF Chip Capacitor
L1 — 8.2 nH Chip Inductor
L2 — 270 nH Chip Inductor

L3 — 150 nH Chip Inductor
RF Connectors — SMA Type
Board Material — Epoxy/Glass ε_r = 4.5,
Dielectric Thickness = 0.014″ (0.36 mm)

Figure 1. Test Circuit Configuration

If the carrier is *phase coherent* with the incoming modulated carrier signal (that is, there is no frequency or phase difference between them, $\theta = 0°$), then the output is proportional to $a(t)$ and perfect detection is achieved.

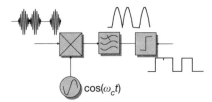

If $\theta = 90°$, however, then $\cos(90°) = 0$ and no output is obtained! It is thus essential to ensure that the carrier oscillator in the receive modem unit is in some way *phase-locked* to the carrier oscillator in the transmitter modem.

Although coherent detection appears much more complicated than non-coherent detection, it is able to recover the data signal more accurately in the presence of noise.

EXAMPLE 5.2

A coherent ASK demodulator has a 5° error in its locally generated carrier reference. What will be the degradation in noise power immunity compared with an ideal demodulator?

Solution

The **voltage** output of the mixer used to compare the incoming symbol $a(t) \cdot \cos(\omega_c t)$ with the reference $\cos(\omega_c t + 5°)$ will be reduced by a factor $\cos(5°)$ from its maximum value as a result of the carrier phase error.

This in turn equates to a reduction in symbol **energy** at the input to the receiver of $\cos^2(5°) = 0.9924$ (symbol energy is proportional to power × symbol length or voltage² × symbol length).

The noise components passing through the mixer will also be affected by the phase error in the carrier reference, but since the noise vectors are assumed to be randomly distributed through 360°, the carrier phase error will reduce the effect of some noise vectors and enhance others with the net effect that the **average noise voltage** at the mixer output will remain unchanged.

Hence it is only the symbol energy that is truly affected by carrier reference phase error and not the noise power. This means that the effective received symbol energy to noise power density will be reduced by a factor $1/\cos^2(5°) = 1.0076$, or 0.0033 dB, as a result of the phase error.

Coherent detection vs non-coherent detection

In order to understand why coherent detection gives a better noise performance than non-coherent detection, it is helpful to view the 'phasor or

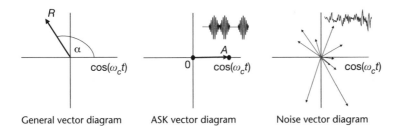

General vector diagram   ASK vector diagram   Noise vector diagram

vector diagram' representation of ASK. A vector diagram maps the amplitude of a signal by the length of the line on the vector diagram, and maps the instantaneous phase by the angle of the line with respect to a horizontal reference frequency and phase (for data modulation, this reference is usually the carrier signal $\cos \omega_c t$). Thus for ASK, we can map the two symbol states (carrier on or off) as two vectors on this diagram, one with length zero corresponding to the carrier-off state, and one with length A in line with the carrier reference, corresponding to the carrier-on state.

We can also map noise by recognizing that a noise signal can be written as a vector $n(t) \cdot \cos[\omega_c t + \theta(t)]$ where $n(t)$ is a time varying amplitude and $\theta(t)$ is a time varying phase. If we were to plot snapshots of the noise on the vector diagram, we would find that they were randomly distributed throughout all four quadrants of the vector space as shown.

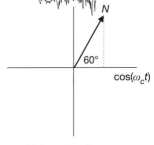

Noise vector diagram

Let us now consider the case of detecting the ASK signal in the presence of noise. For simplicity we will assume that the carrier is in the 'off' state and that we have a specific noise component of length N and phase $60°$.

The non-coherent detector, which is performing amplitude detection, is simply measuring the *length* of the composite (ASK + Noise) vector regardless of the vector phase. It would thus produce an output voltage proportional to N, the noise vector length.

The coherent detector, on the other hand, acts by mixing the incoming signal with the reference carrier $\cos \omega_c t$. The result is that the voltage at the detector output due to the noise is reduced by a factor $\cos(60°) = 0.5$ and is thus proportional to $N/2$. If the noise vector happens to be in phase with the carrier reference, there is no reduction in noise, while if the noise vector is $90°$ out of phase, the noise is reduced to zero. On average, the coherent detection method reduces the noise voltage out of the detector by a factor of $\sqrt{2}$ and the noise power by 2. In other words, coherent detection of ASK can tolerate 3 dB more noise than non-coherent ASK for the same likelihood of detection error.

Carrier recovery for ASK

Clearly it is beneficial to use coherent detection of ASK if possible, but first the problem of how to derive a frequency and phase-coherent carrier reference in the receiver must be addressed. Obtaining sufficiently stable free-running oscillators in both TX and RX units is not possible (see page 86) and some means of recovering the carrier frequency and phase from the incoming data signal is needed.

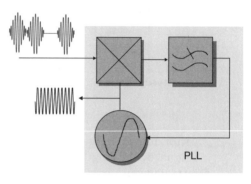

PLL

One method that we have already encountered (see page 97) is to send a reference signal alongside the data signal from which the source carrier frequency and phase can be measured. Here we shall consider an alternative technique that recovers the carrier from the modulated data signal itself. By locking an oscillator to the phase of the incoming carrier when a carrier-on symbol is sent, and holding this oscillator phase when the carrier is off, it is possible to produce the required coherent reference. A technique that is well suited to this task is the phase-locked loop (PLL) (see Section 5.3). Unfortunately the carrier reference is not perfect, as the carrier-on symbols are corrupted by incoming noise. However, by averaging the carrier over several on-symbols (equivalent to a narrow bandwidth in the PLL system), a cleaner reference can be obtained.

Matched filtering for ASK

In Chapter 3, the concept of matched filtering of baseband data signals for optimizing the signal to noise ratio at the output of a data receiver was discussed. Exactly the same approach is applicable to bandpass modulation detection and in fact the bandpass case reverts to the baseband case for ASK if coherent detection is used to 'demodulate' the ASK signal and hence recover the source baseband data stream.

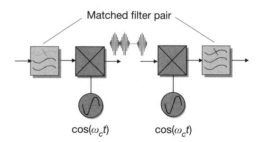

Matched filter pair

$\cos(\omega_c t)$ $\cos(\omega_c t)$

A matched filter pair such as the root raised cosine filters (see Section 3.4) can thus be used to shape the source and received baseband data symbols in ASK, and in fact this is a very common approach to achieving matched filter detection in most bandpass data modems.

Symbol timing recovery

The discussion in Chapter 3 on pulse shaping for minimum intersymbol interference highlighted the need for accurate timing of the sampling point within each symbol. A common symbol timing circuit is the *early-late gate* synchronizer. This circuit works on the basis that the optimum point to sample the signal at the output of a matched filter detector is when the signal is at its maximum.

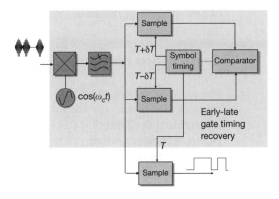

The early-late gate approach uses two such detectors, one fed with a slightly advanced timing reference and one fed with a slightly retarded timing reference. The outputs of the two detectors are then periodically compared to see which is the larger. The timing is then advanced in favour of the detector with the larger output in the expectation that it will get bigger. Eventually, the timing will be advanced too far and the detector output begins to fall. The equilibrium point occurs when both detector outputs are equal (one falling away from the peak and one rising towards it), and the optimum sampling point is then known to lie midway between the advanced and retarded references. This optimum timing signal is passed to a third data detector.

Two excellent references on timing and carrier recovery circuits are Lindsey and Simon (1972) and Gardner (1966).

BER performance of ASK

We saw in Chapter 3 that the performance of a digital communications system is at the simplest level presented as a probability of bit error, or a probability of symbol error, as a function of the received E_b/N_0 ratio. For a binary modulation system, symbol and bit error probability are the same.

Binary ASK effectively uses a unipolar baseband modulation source and the performance for matched filter detection has already been derived in Chapter 3. The BER performance for coherent and non-coherent ASK is shown here for an additive white Gaussian noise limited channel.

The E_b/N_0 value is for the *average symbol power*, which is 3 dB less than the *peak symbol power* for ASK

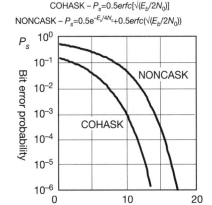

(the carrier is off for approximately half of the transmitted symbols). If performance is to be gauged in terms of peak power, then ASK suffers a penalty of 3 dB over the performance shown here.

Constellation diagrams

The *constellation diagram* is very similar to the vector diagram introduced earlier in the section, and is a method of representing the symbol states in a carrier modulated bandpass modem in terms of their amplitude and phase. Typically, the horizontal axis is taken as a reference for symbols that are *in-phase* with the carrier $\cos(\omega_c t)$, and the vertical axis represents the quadrature carrier component, $\sin(\omega_c t)$. With binary ASK, there are just two symbol states to map onto the constellation space: $a(t) = 0$ (no carrier amplitude, giving a point at the origin), and $a(t) = A \cdot \cos(\omega_c t)$ (giving a point on the positive horizontal axis at a distance A from the origin).

Multi-level ASK systems (see Section 6.2) are represented by adding appropriate points on the constellation diagram. An 8-ary ASK system is depicted here.

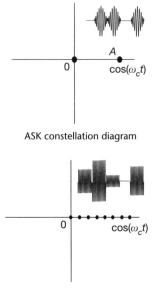

ASK constellation diagram

8-ary ASK vector diagram

EXAMPLE 5.3

Sketch the constellation diagram for 8-ary ASK with a bipolar modulating signal and with a carrier frequency of $\cos(\omega_c t + 45°)$. What would be the output from a non-coherent detector for this type of ASK waveform?

Solution

A constellation diagram is usually drawn with symbols having a phase $\cos(\omega_c t + 0°)$ on the horizontal axis, and those with phase $\cos(\omega_c t + 90°)$ on the vertical axis. For the carrier in this example, the symbols must lie on a line at 45° to the horizontal axis as shown below.

The bipolar nature of the input modulating signal means that the amplitudes of the ASK symbols will have both positive and negative values. A negative value amplitude for an ASK symbol simply means that the carrier phase is inverted for these symbols (that is, they appear in the opposite quadrant of the constellation diagram).

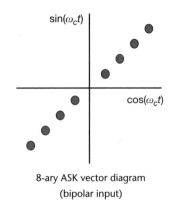

8-ary ASK vector diagram
(bipolar input)

A bipolar ASK signal cannot be detected properly with an envelope detector because it will not pick up the carrier inversion. All negative amplitude symbols will thus be decoded incorrectly as the equivalent positive amplitude symbols.

5.3 Frequency Shift Keying (FSK)

FSK waveforms

Frequency Shift Keying (FSK) has until recent years been the most widely used form of digital modulation, being simple both to generate and to detect, and also being insensitive to amplitude fluctuations in the channel. FSK conveys the data using distinct carrier frequencies to represent symbol states. An important property of FSK is that the amplitude of the modulated wave is constant.

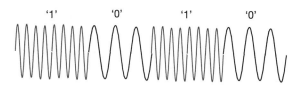

Consider the case of unfiltered binary FSK shown here. This waveform can be viewed as two separate ASK symbol streams summed prior to transmission.

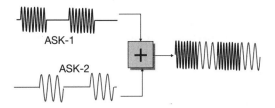

FSK generation

FSK can be generated by switching between distinct frequency sources; however, it is likely that there will be discrete phase jumps between the symbol states at the switching time. Any phase discontinuity at the symbol boundary will result in a much greater prominence of high frequency terms in the spectrum, implying a wider bandwidth for transmission.

Alternatively, FSK can be realized by applying the data signal as a control voltage to a *voltage controlled oscillator* (*VCO*) (see in-depth section). Here the

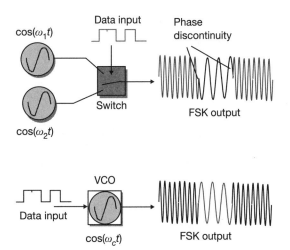

phase transition between consecutive symbol states is guaranteed to be smooth (continuous). FSK with no phase discontinuity between symbols is known as a *Continuous Phase Frequency Shift Keying (CPFSK)* format.

IN DEPTH

Voltage controlled oscillators

A voltage controlled oscillator is a device which produces a sinusoidal (sometimes square wave) output whose frequency is a function of the applied control voltage.

Below is a plot for a typical VCO operating with a frequency near 10 MHz, showing output frequency against control voltage. Ideally, the VCO would give a frequency change that is linearly proportional to the applied voltage as shown by the straight line. In practice, however, most practical VCOs have a response which is characteristically 'S' shaped with a near-linear central portion and greater deviation at the extremes of the frequency range.

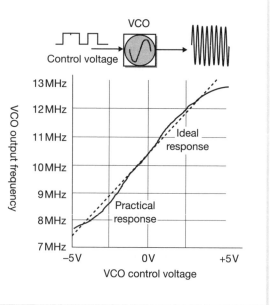

The frequency varying element of a high frequency VCO is usually a 'varactor diode', whose capacitance varies as a function of applied voltage. This is then used as part of an L-C tuned circuit in the feedback path of an amplifier, thus forming a tuneable (voltage controlled) oscillator. Often the inductance is realized using a quartz crystal which has a very high Q and hence produces an accurate and repeatable frequency of oscillation. These devices are usually called a voltage controlled crystal oscillator or VCXO.

The vector modulator

A third method of FSK generation uses a *vector or quadrature modulator* as shown here. The basic vector modulator was introduced in Chapter 1, and in fact can be used to generate any modulation format with the appropriate choice of in-phase and quadrature drive signals. It operates on the principle that any modulation vector can be realized by summing appropriate amounts of an in-phase ($\cos \omega_c t$) and quadrature ($\sin \omega_c t$) version of the carrier signal (see in-depth).

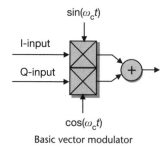

Basic vector modulator

To generate ASK, for example, the I-input to the vector modulator would be fed with the data stream and the Q-input tied to zero. FSK requires the generation of two symbols, one at a frequency $(\omega_c + \omega_1)$ and one at a frequency $(\omega_c - \omega_1)$, for example. In order to generate a frequency shift of $+\omega_1$ at the output of the vector modulator, the I and Q inputs need to be fed with $-\cos\omega_1$ and $\sin\omega_1$ respectively. To generate a shift of $-\omega_1$ requires inputs of $\cos\omega_1$ and $\sin\omega_1$. This approach is now frequently used to generate some of the more elaborate filtered CPFSK formats described later in this chapter – particularly in cellular handsets.

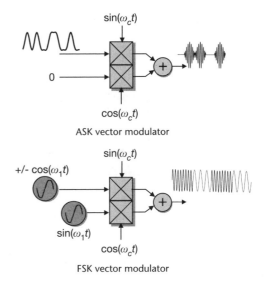

ASK vector modulator

FSK vector modulator

Spectrum of FSK

The spectrum of the FSK signal is not as easy to derive as that for ASK because the FSK generation process is non-linear. An approximation can be obtained by plotting the spectra for two ASK streams centred on the respective carrier frequencies.

Clearly, the overall bandwidth occupied by the FSK signal depends on the separation between the frequencies representing the symbol states.

An FSK system using continuous phase transitions will have much lower side-lobe energy than the discontinuous case.

Spectrum of CPFSK

The spectrum of the CPFSK signal also changes as a function of the frequency spacing between the two symbol states and how the phase trajectory is controlled when changing from one frequency to another. A detailed mathematical description (see in-depth section) of the spectral response for CPFSK can be found in Proakis (1989), and we shall consider only two special cases known as *Sunde's FSK* and *Minimum Shift Keying* (MSK).

Sunde's FSK arises when the spacing between the two symbol frequencies is made exactly equal to the symbol rate. For this case, the spectrum uniquely

IN DEPTH

Example of integrated quadrature (vector) modulator. Data sheet courtesy of Motorola.

MOTOROLA
SEMICONDUCTOR TECHNICAL DATA

Order this document
by MRFIC0001/D

Advance Information
The MRFIC Line
Quadrature Modulator

MRFIC0001

The MRFIC0001 is an integrated Quadrature Modulator designed for operation in the 50 to 260 MHz frequency range. The design utilizes Motorola's advanced MOSAIC 3 silicon bipolar RF process to yield superior performance in a cost effective monolithic device. Applications include DQPSK for PDC, NADC, and PHS; GMSK for GSM and DCS1800; and QPSK for CATV.

- Linear I/Q Ports
- On Chip LO Phase Shifter
- I/Q Phase Imbalance = 2 degrees (Typ)
- I/Q Amplitude Imbalance = 0.3 dB (Typ)
- Gain Control = 30 dB (Typ)
- Single Source Low Operating Supply Voltage
- Low Power Consumption
- Low–Cost, Low Profile Plastic TSSOP Package
- Order MRFIC0001R2 for Tape and Reel.
  R2 Suffix = 2,500 Units per 16 mm, 13 inch Reel.
- Device Marking = M001

QUADRATURE
MODULATOR
INTEGRATED CIRCUIT

CASE 948D–03
(TSSOP–20)

ABSOLUTE MAXIMUM RATINGS (T_A = 25 °C unless otherwise noted)

Parameter	Symbol	Value	Unit
Supply Voltage	V_{CC}	6.5	Vdc
Control Voltages	TX EN, VCNTL	6.5	Vdc
LO Input Power	P_{LO}	0.0	dBm
Differential I/Q Input Voltage	V_D	2.0	V_{pp}
I, I, Q, and Q DC Bias Voltage	V_B	2.0	Vdc
Ambient Operating Temperature	T_A	–30 to +85	°C
Storage Temperature	T_{stg}	–65 to +125	°C

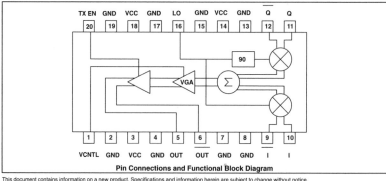

Pin Connections and Functional Block Diagram

This document contains information on a new product. Specifications and information herein are subject to change without notice.

REV 4

 MOTOROLA

RECOMMENDED OPERATING CONDITIONS

Parameter	Symbol	Value	Unit
Supply Voltage	V_{CC}	2.7 to 5.5	Vdc
LO Input Power	P_{LO}	−10	dBm
LO Frequency	f_{LO}	50 to 260	MHz
Differential I/Q Input Voltage	V_D	0 to 1.0	Vdc
I, Ī, Q, and Q̄ DC Bias Voltage	V_B	1.5 to 1.7	Vdc
Variable Gain Amplifier Control Voltage	V_{cntl}	0 to V_{CC}	Vdc
Transmit Enable Low Voltage	TX EN	0 to 0.2	Vdc
Transmit Enable High Voltage	TX EN	V_{CC} − 0.2 to V_{CC}	Vdc

ELECTRICAL CHARACTERISTICS (V_{CC} = 3.0 V, TX EN = 3.0 V, V_{cntl} = 0.0 V, V_D = 0.8 V_{PP}, V_B = 1.6 V, P_{LO} = −10 dBm, f_{LO} = 248 MHz, f_D = 100 kHz, T_A = 25 C unless otherwise noted)

Characteristic	Min	Typ	Max	Unit
Supply Current	–	10	12	mA
Standby Current (TX EN = 0.0V)	–	40	100	μA
Single Sideband Output Power Level	−15	−13	–	dBm
Single Sideband Output Power 1dB Compression Point	–	−10	–	dBm
LO Leakage[2]	–	−55	−45	dBm
Undesired Sideband Level	–	−35	−30	dBc
Output Level Dynamic Range (V_{cntl} = 0 to 2.2V)[2]	–	30	–	dB
Turn–on/off Time	–	2	–	μs

I/Q Data
    Input 3dB Bandwidth
    Amplitude Imbalance
    Phase Imbalance
	–
	–
	–
		5
		0.3
		2
			–
			–
			–
				MHz
				dB
				degree

(1) All electrical characteristics measured in test circuit schematic shown in Figure 1.
   V_B is the bias voltage on the input data ports.
   V_D is the sinusoidal differential voltage on the input data ports when testing the part in a single sideband mode.
   Above power levels are the single–ended output power.
(2) LO leakage power is unaffected by V_{cntl} setting.

EVALUATION BOARDS

Evaluation boards are available for RF Monolithic Integrated Circuits by adding a "TF" suffix to the device type. For a complete list of currently available boards and ones in development for newly introduced product, please contact your local Motorola Distributor or Sales Office.

contains two discrete spectral lines at the two symbol frequencies, in addition to a broad spectral spread. These spectral lines can be used in a coherent FSK detector as the source of the carrier references, often extracted using a phase-locked loop (see below).

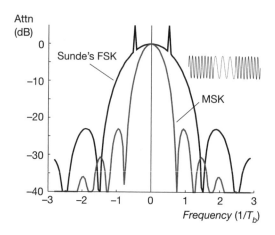

Minimum Shift Keying employs a symbol spacing equal to one half of the symbol rate and produces a smooth spectrum with narrow main lobe and rapidly reducing side-lobe energy. This narrow symbol spacing means that MSK can be more spectrally efficient than binary ASK and PSK, and in fact approaches the performance of Quadrature Phase Shift Keying (QPSK) systems (see Section 6.4). The price to be paid for this excellent performance is increased complexity in the generation and detection process compared with Sunde's FSK, for example.

IN DEPTH

Spectrum of Frequency Shift Keying waveforms

The spectrum of FSK depends on a number of factors, including whether the transition between symbol states has a continuous phase or discontinuous phase, whether the data waveform driving the modulator (typically a VCO) is shaped by filtering, and the frequency separation between symbol states. A good insight into the whole topic is given by Lucky *et al.* (1968).

Only two special cases will be considered here – binary FSK with a frequency separation of $1/T_b$ and no pulse shaping (this is often called Sunde's FSK), and binary Continuous Phase FSK with frequency separation of $0.5 \times 1/T_b$ and no pulse shaping (usually called Minimum Shift Keying).

Power spectral density for Sunde's FSK

$$G(f)_{\text{Sunde's-FSK}} = \frac{1}{4}[\delta(f - 0.5 \times 1/T_b) + \delta(f + 0.5 \times 1/T_b)] + P(f)$$

where

$$P(f) = \frac{4T_b}{\pi^2}\left[\frac{\cos(\pi f T_b)}{4f^2 T_b^2 - 1}\right]^2$$

The characteristic of Sunde's FSK is that it has two discrete components in the spectrum at the two symbol frequencies. These are of great benefit when recovering a carrier component for coherent detection.

Power spectral density of Minimum Shift Keying (MSK)

$$G(f)_{MSK} = \frac{16T_b}{\pi^2} \left[\frac{\cos(2\pi f T_b)}{16f^2 T_b^2 - 1} \right]^2$$

Unlike Sunde's FSK, this spectrum has no discrete components, and a much narrower main lobe as would be expected with the narrower frequency separation between symbols. Comparing this result with the power spectral density of QPSK, given by:

$$G(f)_{QPSK(unfiltered)} = T_b \sin c^2 (2\pi T_b f)$$

we can see that the side-lobe energy for MSK falls off much more quickly than for QPSK, with MSK having a slightly wider main lobe. As both formats offer a potential bandwidth efficiency approaching 2 bits/second/Hz, MSK is often the preferred modulation choice in systems where the constant envelope property of the FSK family is important. Where filtered QPSK with its associated amplitude fluctuations can be tolerated, this will give the ultimate minimum bandwidth solution.

Filtered FSK

As with ASK modulation, it is possible to control the spectral occupancy of CPFSK with a pulse-shaping filter prior to the modulator. In the case of FSK, however, there is not a one-to-one mapping between the pulse-shaped spectrum and the modulated spectrum owing to the non-linear modulation process (for example, VCO).

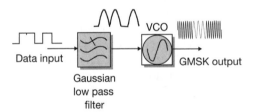

The filter commonly used is an unusual low pass filter called a *Gaussian filter*, which is particularly well suited to achieving low spectral occupancy outside the FSK main lobe. A modern example of the use of this type of FSK, commonly termed *Gaussian Minimum Shift Keying*, is within the European and North American (GSM-based) digital cellular systems.

Non-coherent FSK detection

One of the simplest ways of detecting binary FSK is to pass the signal through two bandpass filters tuned to the two signalling frequencies and detect which has the larger output averaged over a symbol period. This is in essence a non-coherent envelope detector for the equivalent two ASK streams with a comparator at the output. Because it takes no account of the phase of the respective symbols, this method, as expected from our discussions on ASK detection (see Section 5.2), will not perform as well as coherent FSK detection systems.

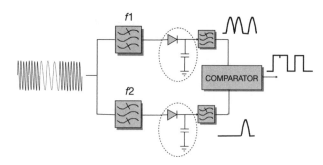

There are several alternative methods for discriminating between the incoming frequencies. One simple digital method involves counting the zero-crossings of the carrier during a symbol and hence directly estimating the frequency on a symbol-by-symbol basis. A third, often used method involves a phase-locked loop which is described next.

Non-coherent PLL-based FSK detection

The *phase-locked loop* (PLL) is frequently used in symbol and carrier recovery circuits for digital communications systems. The basic operation of the PLL is described here in connection with its use as an FSK detector. For a more detailed discussion of PLL systems there is an excellent book by Gardner (1966).

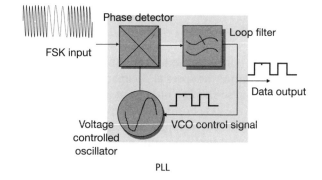

The PLL consists of three building blocks: a *voltage controlled oscillator* whose output frequency is proportional to the input voltage, a *phase detector* (often implemented using a multiplier or exclusive-or gate) which produces a voltage output proportional to the phase difference between the two inputs, and a *loop filter* which is used to control the dynamics of the *feedback* circuit.

The PLL acts by comparing the phase of the input signal with that of the VCO and using the voltage generated by the phase difference to alter the frequency and phase of the VCO to 'match' that of the input. The system reaches a stable steady state when the average output from the phase detector is zero, implying that the VCO is 'phase-locked' to the input signal. (With mixer-based phase detectors, there is a 90° phase difference between input and VCO in the phase-locked state.) Because the VCO control voltage must change in order for the PLL to track and lock onto a new input frequency, it provides a direct measure of the input signal frequency for each symbol in the FSK stream and hence acts as a first-class detector.

E EXAMPLE 5.4

A digital radio system uses binary FSK for data transmission, with the two symbol frequencies at +1200 Hz and −1200 Hz with respect to the channel centre. The received signal is subject to a Doppler shift of +100 Hz owing to the receiver motion. Sketch the output of a PLL FSK detector for a 1,0,1,0,1,0, ... data stream assuming there is no pulse shaping. How can the problem of Doppler shift be overcome in a digital FSK radio system?

Solution

The effect of the Doppler shift on the receiver signal is to make the symbols appear at frequencies of +1300 Hz and −1100 Hz with respect to the notional centre frequency of the PLL detector. The result is that the PLL output will have a dc-bias voltage superimposed on the recovered data signal as shown below, proportional to the Doppler offset.

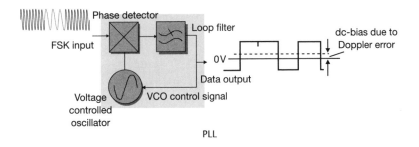

To eliminate the Doppler shift the output of the PLL detector can be ac-coupled; however, this would also affect any low frequency content in the data signal itself. A coding scheme such as Manchester encoding (see Section 4.4) could be used here to remove any low frequency content in the data signal.

Coherent FSK detection

Coherent detection of FSK is very similar to that for ASK but in this case there are two detectors tuned to the two carrier frequencies. As for ASK, coherent detection and matched filtering minimize the effects of noise in the receiver.

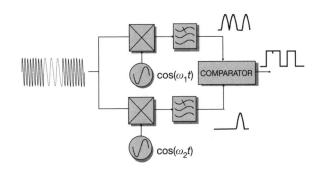

Recovery of the carrier references in the coherent receiver is made simple if the frequency spacing between the symbols is made equal to the symbol rate (Sunde's FSK), as the modulated spectrum contains discrete spectral lines at the carrier frequencies. The drawback of using Sunde's FSK is that the bandwidth of the FSK signal is approximately 1.5 to 2 times that of an optimally filtered ASK or PSK binary signal.

BER performance for FSK

The *theoretical* performance for coherent and non-coherent FSK is shown here for an Additive White Gaussian Noise limited channel. If the FSK symbol frequencies are chosen to be orthogonal (see Section 6.3) such that the two coherent detectors can in essence operate independently of each other, then the BER performance for coherent FSK is identical to that for coherent ASK.

It can be seen that the E_b/N_0 penalty of the simpler non-coherent detection method is only about 1 dB at practical bit error rates. As a result, the simpler, non-coherent FSK forms the basis of many low-end (for example, 1200 bps) telephone and radio modems in the market-place. Interestingly, the non-coherent performance of FSK is not nearly as bad as that for ASK.

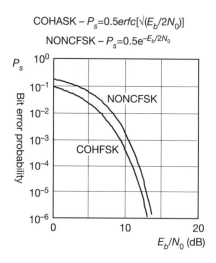

COHASK – P_s=0.5$erfc$[√($E_b/2N_0$)]

NONCFSK – P_s=0.5$e^{-E_b/2N_0}$

Advantages of FSK

- FSK is a *constant envelope modulation,* and hence insensitive to amplitude (gain) variations in the channel and compatible with non-linear transmitter and receiver systems.

- The detection of FSK can be based on *relative frequency changes* between symbol states and thus does not require absolute frequency accuracy in the channel. (FSK is thus relatively tolerant of local oscillator drift and Doppler shift (see Section 4.2).)

Disadvantages of FSK

- FSK is slightly less bandwidth efficient than ASK or PSK (excluding MSK implementation).

- The bit/symbol error rate performance of FSK is worse than for PSK.

5.4 Phase Shift Keying (PSK)

Principle of PSK

With *Phase Shift Keying (PSK)*, the information is contained in the instantaneous phase of the modulated carrier. Usually this phase is imposed and measured with respect to a fixed carrier of known phase – *coherent PSK*. For binary PSK, phase states of 0° and 180° are used.

It is also possible to transmit data encoded as the phase change (phase difference) between consecutive symbols. This method is classified as *Differentially Coherent PSK*.

There is no non-coherent detection equivalent for PSK.

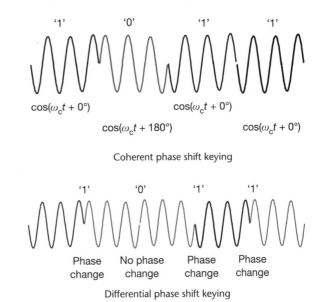

$\cos(\omega_c t + 0°)$ $\cos(\omega_c t + 0°)$

$\cos(\omega_c t + 180°)$ $\cos(\omega_c t + 0°)$

Coherent phase shift keying

'1'   '0'   '1'   '1'

Phase change   No phase change   Phase change   Phase change

Differential phase shift keying

Spectral occupancy for PSK

The bandwidth of a binary PSK signal is identical to that of binary ASK, assuming the same degree of pulse shaping. In fact, BPSK can be viewed as an ASK signal with the carrier amplitudes as $+A$ and $-A$ (rather than $+A$ and 0 for ASK).

If the phase changes are abrupt at the symbol boundaries, then, just like FSK, the occupied bandwidth will be much larger than for smooth transitions between phase states, implying the need for shaping of the modulation waveform.

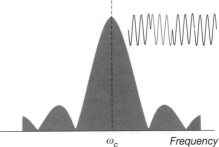

ω_c *Frequency*

BPSK spectrum

PSK generation

The simplest means of realizing unfiltered binary PSK is to switch the sign of the carrier using the data signal, causing a 0° or 180° phase shift. Just as for ASK, this method of generation is not well suited to obtaining a Nyquist filtered waveform owing to the

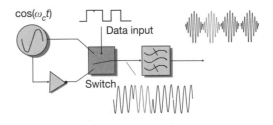

$\cos(\omega_c t)$

Data input

Switch

difficulty in implementing bandpass high
frequency, high Q filters (see Section 3.2).

If filtering is required, then linear
multiplication must be employed, allowing
the data stream to be pre-shaped at baseband
prior to the modulation process. Because the
modulation process is linear, the baseband
filter shape is imposed directly onto the
bandpass modulation signal.

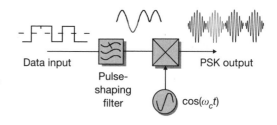

Data input   Pulse-shaping filter $\cos(\omega_c t)$ PSK output

The effect of filtering on the PSK waveform

While an unfiltered PSK signal has a constant envelope, the introduction of
filtering to constrain the modulation bandwidth reintroduces an envelope
variation to the PSK signal. The degree of envelope modulation introduced
is a function of the severity of the pulse
shaping employed.

Shown here are examples of PSK
waveforms for various values of root raised
cosine filter roll-off. As expected, the smaller
the value of α, the sharper the filter and the
higher the peaks of the PSK signal. (The *root
raised cosine filter* is used here, as it is the
peak power rating of the *transmitter* that is
usually of most concern to designers.)

Further information on the peak to mean
power ratios for M-ary PSK and QAM signals
can be found in Chapter 6.

Root RC filtered PSK, α=0.1

Root RC filtered PSK, α=1.0

Unfiltered PSK

Detection of PSK

There is no *non-coherent* equivalent
detection process for PSK, and
various forms of coherent detection
must be employed. The ideal detector
thus requires perfect knowledge of the
unmodulated carrier phase at the
receiver.

As with ASK, any phase error θ of
the locally generated carrier reference

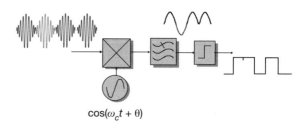

$\cos(\omega_c t + \theta)$

reduces the signal voltage at the output of the detector by a factor $\cos(\theta)$. This
in turn (see example) degrades the E_s/N_0 performance of the detector by a

factor $\cos^2(\theta)$. Thus we need zero phase error for optimum detection and must revisit the whole area of *carrier recovery*. Note that if the phase error reaches 90°, the output falls to zero!

E EXAMPLE 5.5

A coherent PSK demodulator uses a carrier recovery circuit which introduces a time delay in the recovered reference equal to 7% of the carrier period. What will be the degradation in noise power immunity for this system compared with an ideal demodulator?

Solution

A time delay of 7% of the carrier period results in a phase error in the recovered carrier of 7% or 360° = 25.2°.

The **voltage** output of the mixer used to compare the incoming symbol $a(t) \cdot \cos(\omega_c t)$ with the reference $\cos(\omega_c t + 25.2°)$ will be reduced by a factor $\cos(25.2°)$ from its maximum value as a result of the carrier phase error.

This in turn equates to a reduction in symbol **energy** at the input to the receiver of $\cos^2(25.2°)$. (Symbol energy is proportional to power × symbol length or voltage^2 × symbol length.)

The noise components passing through the mixer will also be affected by the phase error in the carrier reference, but since the noise vectors are assumed to be randomly distributed through 360°, the carrier phase error will reduce the effect of some noise vectors and enhance others with the net effect that the **average noise voltage** at the mixer output will remain unchanged. Hence it is only the symbol energy that is truly affected by carrier reference phase error and not the noise power. This means that the effective received symbol energy to noise power density will be reduced by a factor $1/\cos^2(25.2°) = 1.22$, or 0.87 dB, as a result of the phase error.

Carrier recovery for coherent PSK

In order to ensure that the local carrier phase is approximately 0°, it is necessary either to transmit a *carrier phase reference* alongside the data signal, or to derive the reference from the incoming data signal. The concept of a separate reference signal has been discussed already in Chapter 4 as a means of detecting and correcting for local oscillator error and Doppler shift in mobile radio systems.

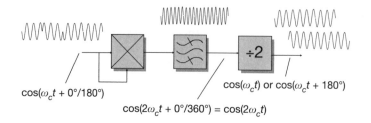

$\cos(\omega_c t + 0°/180°)$

$\cos(2\omega_c t + 0°/360°) = \cos(2\omega_c t)$

$\cos(\omega_c t)$ or $\cos(\omega_c t + 180°)$

A data-derived reference for BPSK can be implemented by realizing that if the binary PSK signal is squared, the 0° and 180° phase states can be forced to be modulo 2π, thereby removing the modulation. The process of squaring also doubles the frequency of the carrier component. This twice frequency term requires filtering (usually with a phase-locked loop), to remove channel noise, and then the frequency must be halved to yield the required coherent carrier term.

For data systems using N different phase symbol states ($N = 2$ for binary PSK), an Nth order non-linearity must be used to force the phase modulation to be modulo 2π. The remainder of the carrier recovery process remains the same except that a frequency 'divide by N' circuit is needed to obtain the correct carrier frequency.

For practical filtered PSK, the squared signal contains an envelope modulation component which manifests itself as additional symmetrical frequency components around the twice carrier term. Fortunately a PLL-based filter is insensitive to envelope modulation and this will not significantly affect the performance of the carrier recovery circuit.

The filtering that is applied to the twice carrier term to reduce the effects of channel noise is,

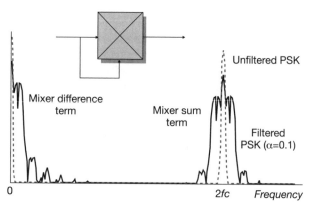

Output of squarer for filtered vs unfiltered PSK

however, very important in reducing the amount of *phase jitter* on the recovered carrier reference. There are some applications of digital communications, such as cellular radio, where it is not possible to employ a very narrow filter in the carrier recovery circuit. This is owing to the frequency uncertainty in the location of the twice carrier term caused by local oscillator error and, more directly, variable amounts of Doppler shift (see Section 4.2) resulting from the user's motion. In these cases, the carrier recovery process can often have significant residual 'phase jitter' which degrades the performance for the expected ideal coherent PSK detector.

The Costas loop

A commonly used variant of the squaring-based carrier recovery method is the *Costas loop* (Costas, 1956).

The Costas loop is in essence two phase-locked loops operating in parallel, with a common VCO giving quadrature outputs to each loop. The squaring process necessary to make the PSK modulation modulo 2π is inherent within the Costas loop by virtue of the third mixer.

There are two main advantages to the Costas loop. Firstly, it does not implicitly generate a twice carrier frequency component and thus does away with the need for a 'divide by two' circuit. Secondly, it performs the required coherent data detection in one of the branches of the dual PLL system, eliminating the need for extra detection circuitry. (A separate matched filter (see Section 3.5) is used for data recovery as the filtering within the Costas loop is usually made very narrow in order to achieve good noise averaging of the coherent reference.)

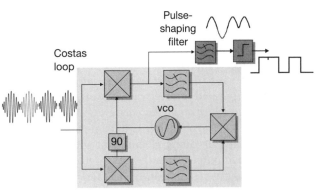

Phase ambiguity in PSK carrier recovery

The 'squarer'-based form of carrier recovery looks ideal, but unfortunately it suffers from one significant drawback – the process of halving the frequency of the twice carrier term introduces a 180° *phase ambiguity* into the carrier reference.

E.g. 11110000

Consider the case of a 1,0,1,0,1,0 ... filtered data stream entering the squaring circuit. The output will be a twice carrier term, with zero-crossings at twice the rate of the input. Feeding this signal to a 'divide by two' circuit, it is clear that the divider logic can be triggered by either of the zero-crossings in the twice carrier waveform, and can have no knowledge of which one relates to the correct zero-crossing (and hence phase) of the input. The result is that the recovered carrier may be coherent with 0° phase error, or it may be inverted with 180° phase error. Feeding the inverted reference to the coherent detector will result in all of the detected data being inverted!

This problem can be resolved by sending a *training sequence* known to the receiver from which it can deduce that data inversion has occurred and correct accordingly. The training sequence approach does not work well, however, if the channel is frequently interrupted (for example, in a fading mobile environment), resulting in periodic loss of carrier reference. Each time the carrier reference is re-established, an unknown phase ambiguity will again be present, requiring a repeated retransmission of the training sequence. This phase ambiguity problem is also present within the Costas loop even though it does not have an implicit divider circuit.

Differential data encoding

An alternative and more frequently used method of coping with phase ambiguity in the carrier recovery process is to use *differential encoding and decoding* of the input and received data stream. This process is termed *Differentially Encoded Phase Shift Keying (DEPSK)*. The challenge is to come up with an encoding/decoding scheme that will give the same decoded output regardless of whether the received data is inverted or not.

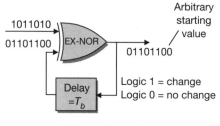

Differential encoder

Shown here is such an encoder based on an exclusive-or gate. This circuit operates by translating the input data into a coded data stream where a logic 1 at the input is coded as a *change of logic state* from the previous coded bit, and a logic 0 at the input as *no change of state* from the previous coded bit. Implementation of the 1-bit delay block can be achieved very simply using a clocked shift register.

This encoding process is very efficient as it does not introduce any extra data bits and hence does not affect the throughput of the data modem.

Differential data decoding

The differential decoding process is equally simple to implement using a second exclusive-or gate and a 1-bit delay. The task is to observe whether the detected data stream changes state over consecutive bits, in which case a logic 1 must have been present in the input. If there is no change of state, a logic 0 must have been sent. This change of state information is unaffected by any data inversion and hence the encoding/decoding process is foolproof against carrier recovery phase ambiguity.

01101100
Delay $=T_b$
EX-NOR
01101100X
1011010X
Arbitrary starting value
Logic 1 = change
Logic 0 = no change
X = don't care

Differential decoder

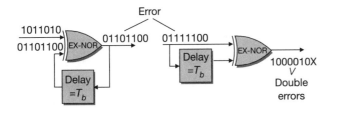

The only drawback with this coding process is that when single bit errors occur in the received data sequence due to noise and so on, and pass through the decoder, they tend to propagate as 'double bit' errors. This is because the decoder is comparing the logic state of the most recently received data bit with the current data bit, and if the previous bit is in error, the next decoded bit will also be in error.

Differential PSK (DPSK)

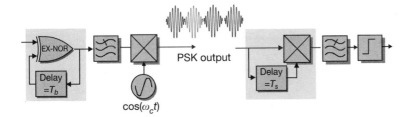

Differential PSK (DPSK) is based on the same 'change of state' encoding/decoding methodology as used in DEPSK, but improves upon it by incorporating the differential decoding task as part of the data demodulation task, and at the same time does away with the need for a 'carrier recovery' mechanism.

The differential encoding block and PSK modulator is common to both DPSK and DEPSK, but the receiver operates by comparing the phase of the current incoming carrier symbol with that of the previous carrier symbol. In this process, it rolls 'coherent detection' and 'differential decoding' into one operation.

Clearly, this detection process is much simpler than that required for true coherent PSK and consequently DPSK is widely used in wired and radio modems for medium-rate signalling (up to 4800 bps). DPSK, however, has a slightly poorer noise immunity than PSK since the phase reference for DPSK is now a *noisy delayed version* of the input signal rather than potentially a well-filtered, virtually noiseless reference from a carrier recovery process.

EXAMPLE 5.6

A DPSK receiver has an error in the delay element equivalent to 10% of the carrier period. Given that the expression for bit error probability in a perfect (no delay error) DPSK receiver is of the form:

$$P_e = 0.5e^{E_b/N_0}$$

where E_b is the energy per bit, and N_0 is the noise power density, what is the percentage increase in transmitter power required to counteract the degradation in receiver performance caused by the delay error?

Solution

The time delay error = 10% of carrier period = 36 degrees of phase error in the notional carrier reference.

The **voltage** output of the mixer used to compare the present symbol $\cos(\omega_c t)$ with the previous symbol $\cos(\omega_c t + T_b + 36°)$ will thus be reduced by a factor $\cos(36°)$ from its maximum value as a result of this timing error.

This in turn equates to a reduction in symbol **energy** at the input to the receiver of $\cos^2(36°)$.

The noise components passing through the mixer will also be affected by the phase error in the carrier reference, but since the noise vectors are assumed to be randomly distributed through 360°, the carrier phase error will reduce the effect of some noise vectors and enhance others, with the net effect that the **average noise voltage** at the mixer output will remain unchanged. Hence it is only the symbol energy that is truly affected by the timing error and not the noise power. The bit error probability can thus be written as:

$$P_e = 0.5e^{E_b \cdot \cos^2(36°)/N_0}$$

This means that the transmitted symbol energy must be increased by a factor $1/\cos^2(36°) = 1.52$, or 1.85 dB, to maintain the same performance as a system with no timing error.

Symbol timing recovery for PSK

The symbol timing recovery problem for all the binary modulation formats discussed thus far (ASK, FSK and PSK) is identical to that for the baseband symbol streams, assuming that the timing recovery is performed on the demodulated, filtered data.

Techniques based on zero-crossing detection, 'squaring' and the early-late gate all feature in modern modem implementations, together with techniques based on dedicated mid-, pre- or post-amble synchronization words.

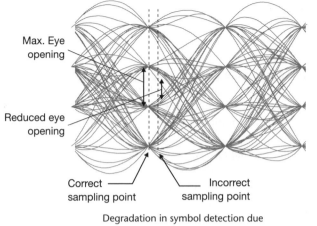

Max. Eye opening

Reduced eye opening

Correct sampling point

Incorrect sampling point

Degradation in symbol detection due to incorrect sample timing

---

E EXAMPLE 5.7

A binary PSK modem uses a preamble data word to achieve symbol synchronization in the receiver. This symbol timing reference, however, drifts between resynchronization bursts so that the timing is in error by 10% when the next synchronization update is received. If the PSK detector uses an integrate and dump filter, and the symbols are not shaped, what will be the worst-case degradation on the effective received symbol energy due to the timing error?

Solution

With reference to the diagram, the 10% timing error in the symbol sampling time will result in the voltage at the output from the integrator when sampled being reduced by 90% of the optimum value. This in turn will result in a fall of 81% (0.9×0.9) in the effective symbol energy, equivalent to a 0.91 dB degradation in E_b/N_0 performance for the modem.

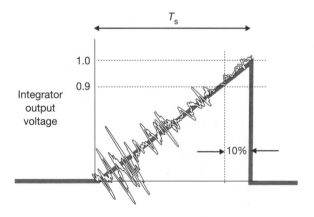

T_s

1.0

0.9

Integrator output voltage

10%

Constellation diagram for PSK

The constellation diagram for binary PSK displays the characteristic of *antipodal* signalling. This means that the symbols used are equal and opposite to each other in the constellation space. Antipodal signalling is a prerequisite for achieving optimum data detection performance in noise, as was discovered when analysing the performance of bipolar baseband signalling in Chapter 3.

If we compare the PSK constellation diagram with that for ASK, we see that the 'antipodal' signalling condition is not satisfied for ASK and hence we do not get the optimum performance in noise. This is directly equivalent to the inferior performance of unipolar signalling compared with bipolar signalling which suffers the same 3 dB performance penalty.

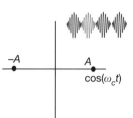

$-A$ A
$\cos(\omega_c t)$

PSK constellation diagram

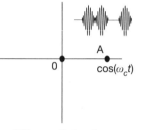

0 A
$\cos(\omega_c t)$

ASK constellation diagram

BER performance for PSK

The *theoretical* performance for coherent and differential PSK is shown here for an additive white Gaussian noise limited channel. The BER probability for coherent PSK is exactly the same as that derived for bipolar baseband transmission in Chapter 3, and the PSK modulation/demodulation process can be viewed as a convenient mechanism for obtaining a bandpass representation of the equivalent baseband source. It should not be forgotten, however, that the baseband-to-bandpass transformation decreases the maximum bandwidth efficiency (see Section 2.2) of the data link from 2 bits/second/Hz to 1 bit/second/Hz for binary signalling in both cases. (We will see in Chapter 6 that Quadrature Phase Shift Keying (QPSK) – a four-symbol PSK scheme – can allow us to get back to an efficiency of 2 bits/second/Hz for the bandpass case without degrading the performance compared with binary PSK.)

COHPSK – $P_s = 0.5 erfc[\sqrt{(E_b/N_0)}]$
DPSK – $P_s = 0.5 e^{-E_b/N_0}$

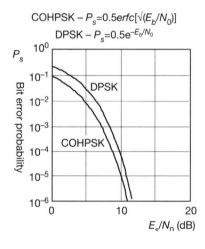

P_s

Bit error probability

DPSK

COHPSK

E_s/N_0 (dB)

It can be seen that there is little performance penalty between fully coherent PSK (assuming a perfect 'jitter-free' reference) and the simpler DPSK implementation. This margin reduces further if differential encoding and decoding is used with coherent PSK to overcome the phase ambiguity within the carrier recovery process which introduces some double bit errors.

EXAMPLE 5.8

A radio receiver is limited in its performance by thermal noise generated in the receiver front-end. The receiver is capable of decoding both coherent binary ASK and binary PSK signals. If the average power rating of the transmitter is constant when it is sending both ASK and PSK data, which modulation format will be able to send data at the fastest rate for a given bit error probability in the receiver, and how much faster can it be?

Solution

The BER equations for coherent ASK and coherent PSK are as follows:

$$P_{e_{ASK}} = 0.5 erfc[E_b/2N_0]^{1/2}$$

$$P_{e_{PSK}} = 0.5 erfc[E_b/N_0]^{1/2}$$

It can be seen that the PSK system can tolerate a noise power density that is twice that for ASK or an E_b value that is half that for ASK.

In this example, the noise power density at the input to the receiver is fixed for both ASK and PSK, and therefore it is possible to halve the E_b value for PSK and still match the ASK performance, allowing the PSK system to send data at twice the rate for the same average transmitter power.

5.5 Comparison of binary modulation schemes

Relative BER performance

Equal average *symbol energy*

The bit error performance for the major binary modulation schemes is shown here, with the PSK family demonstrating optimum performance as would be expected with *antipodal* signalling. Coherent FSK with orthogonal signalling (see Section 6.3) is the next best together with coherent ASK, assuming comparison on an *average symbol* power basis. All of these binary systems exhibit a theoretical maximum bandwidth efficiency of 1 bit/second/Hz.

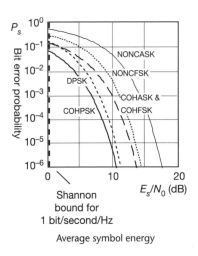

It is interesting to compare these results with the theoretical minimum E_s/N_0 of only 1 (or 0 dB) required for error-free transmission and 1 bit/second/Hz capacity as given by the Shannon–Hartley capacity theorem (see Section 2.4). This demonstrates clearly that implicit within the Shannon–Hartley prediction is significant performance improvement over the basic modem design achieved using message data coding techniques. Consequently, Chapter 7 is given over to the discussion of coding for improved digital communications performance.

Equal peak symbol energy

The bit error performance for the major binary modulation schemes, assuming equal *peak symbol power*, is shown here. This introduces a 3 dB degradation into the performance curves for ASK.

There are many applications where the available power is peak limited, possibly by the power source itself, or by system components – for example, the RF amplifier rating. Other systems, however, are governed by average power, such as average battery drain, or average heat dissipation, and the former set of curves apply.

Although not shown here, the performance of filtered ASK and PSK systems will degrade from the results given owing to the increase in peak to mean power ratio caused by the pulse overshoot at the output of a practical pulse-shaping filter.

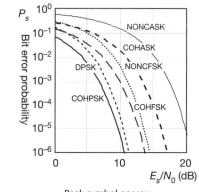

Peak symbol energy

Q QUESTIONS

5.1 An 8-ary ASK scheme is signalling on a channel at a rate of 2400 baud. What is the bandwidth occupied by the modulation signal, assuming ideal brick-wall filtering? If a filter with $\alpha = 0.5$ is used, what will the new occupied bandwidth be?

5.2 A coherent binary ASK data system has a phase error of $25°$ in the recovered carrier reference. What will be the percentage reduction in symbol output voltage from the mixer detector and how much must the input symbol energy be increased to compensate for the loss due to carrier error?

5.3 A binary ASK modem used non-coherent detection. With reference to the BER curves for binary ASK in the text, what is the E_b/N_0 value required to achieve an error probability of less than 1 in 10^2? What is the equivalent performance of a coherent ASK scheme at this E_b/N_0 value?

5.4 A coherent binary ASK modem is found to have a phase error in the recovered carrier of 45°. Will the performance of this modem be improved if non-coherent detection is used instead of the imperfect coherent detection process?

5.5 Draw the constellation diagram for a four-level ASK modulation format using a $\sin(\omega_c t)$ carrier when the modulation input is:

(a) A four-level unipolar signal

(b) A four-level bipolar signal.

5.6 What is the bit error probability for non-coherent binary FSK for an E_b/N_0 value of 10 dB? What approximate E_b/N_0 is required to achieve the same BER performance for coherent FSK and coherent ASK?

5.7 A designer has been asked to build a radio data modem that must be tolerant to a frequency error in the receiver system. He is less concerned about the noise tolerance of the modem. Which modulation format, ASK or FSK, would you recommend for this task?

5.8 A binary PSK modem is designed to work within a bandwidth of 8 kHz. What is the maximum data rate that can be delivered if a raised cosine filter with $\alpha = 1$ is used?

5.9 What is the bandwidth efficiency of a BPSK modem with a pulse-shaping filter with $\alpha = 0.5$?

5.10 A BPSK signal undergoes a Doppler shift of $+70$ Hz in the carrier of the received signal which would normally be at a frequency of 1 MHz. What will be the frequency measured at the output of the squaring circuit in the carrier recovery process?

5.11 When cables are installed in a building, it is not unusual for the engineers to get the connections of the twisted pair reversed. How can a binary signalling scheme be designed to cope with this eventuality and maintain correct polarity data transfer?

5.12 A DPSK transmitter can generate an average power of 1 nW at the input to a receiver which has a noise power density of 0.5×10^{-12} Watts/Hz. If the symbol rate is 100 symbols per second, what is the BER performance for a DPSK decoder in the receiver?

6 Multi-level digital modulation

6.1 Introduction

Thus far, we have only considered bandpass data modulation schemes using binary signalling, and with only one property (*amplitude, frequency* or *phase*) of the carrier wave being altered. We have also seen that for binary ASK, FSK or PSK, the minimum bandwidth required for transmission is at least twice the minimum bandwidth required to pass the baseband binary data stream.

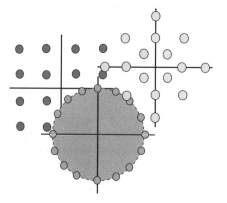

In order to improve on the bandwidth efficiency of bandpass data transmission, we can of course increase the number of symbol states used (except for the case of FSK, where increasing the number of frequencies can increase the occupied bandwidth). As a general rule, however, we know that as the number of symbol states is increased, the tolerance to noise is reduced. We shall see that there are two exceptions to this rule, QPSK and orthogonal M-ary FSK.

M-ary bandpass modulation formats, particularly M-ary Quadrature Amplitude Modulation, are widely used in both wired and wireless digital communications links.

6.2 M-ary Amplitude Shift Keying (M-ary ASK)

Implementation of M-ary ASK

Extending binary ASK to multi-level ASK is a simple concept, and the generation and detection process scales up to requiring multi-level comparison of the recovered envelope signal for either coherent or non-coherent detection. Carrier recovery for M-ary ASK is performed with the same methods as employed with binary ASK.

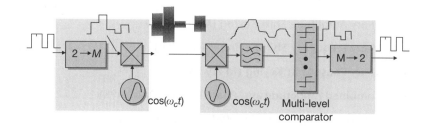

Performance of M-ary ASK

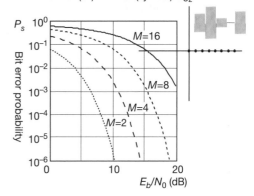

There is no opportunity to exploit orthogonality with M-ary ASK symbol sets and so there is an immediate penalty in BER performance as we move up from the binary system.

The relatively poor BER results for M-ary ASK, coupled with its sensitivity to any gain variations in the channel and the need for reasonable linearity in the transceiver processing, means that there are very few practical examples of ASK other than in its binary form.

M-ASK(symbol): $P_s = [(M-1)/M].erfc[\sqrt{\{3.(E_b/N_0)/(M^2-1)\}}]$

M-ASK(bit) $\approx$ M-ASK(symbol)/$\log_2 M$

P_s

Bit error probability

E_b/N_0 (dB)

6.3 M-ary Frequency Shift Keying (M-ary FSK)

M-ary FSK application

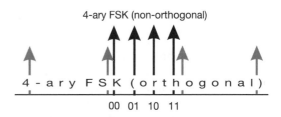

4-ary FSK (non-orthogonal)

4 - a r y   F S K   (o r t h o g o n a l)

00 01 10 11

Unlike M-ary ASK and in contrast to many of the discussions to date, M-ary FSK is very much of interest for *increasing* the noise immunity of the modulation format compared with binary FSK, allowing a designer to achieve reliable data transmission in the presence of high levels of noise. As we shall see shortly, this is only possible by using a set of 'orthogonal symbols', with precisely spaced frequencies which require large amounts of bandwidth. M-ary FSK using orthogonal signalling is one of the few techniques where the modem performance approaches the Shannon bound (see Section 2.4) for minimum E_b/N_0 operation of -1.6 dB.

It is also possible to operate M-ary FSK with 'non-orthogonal' symbol frequencies as we saw with binary FSK. By spacing the frequencies very close together it is possible to squeeze four symbols into the space of two symbols, for example, and hence improve the bandwidth efficiency over BFSK. In this case, the noise immunity for the M-ary FSK system *decreases* compared with the binary system, as the symbol frequencies are no longer orthogonal.

Orthogonal signalling

> Two symbol states $a_i(t)$ and $a_j(t)$ are said to be *orthogonal* over the symbol period T_s if:
>
> $$\int_0^{T_s} a_i(t) \cdot a_j(t) \cdot dt \xrightarrow{i \neq j} 0$$

If the frequencies of M-ary FSK symbols are chosen to be of the form:

$$a(t) = \cos\left(2\pi f_c t + \frac{2\pi m t}{2T_s}\right)$$

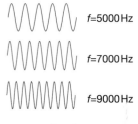

f=5000 Hz

f=7000 Hz

f=9000 Hz

Examples of M-ary FSK orthogonal symbols

where $m = 1, 2, \ldots, M$, then these frequencies are orthogonal over a symbol period.

An orthogonal 8-ary FSK set with a symbol rate of 1200 symbols/sec could thus use frequencies of, say, 1000 Hz, 1600 Hz, 2200 Hz, 2800 Hz, 3400 Hz, 4000 Hz, 4600 Hz and 5200 Hz respectively with the same starting phase.

Properties of orthogonal symbols

The practical interpretation of the orthogonality definition is that if a symbol $a_i(t)$ is mixed with a carrier reference equal to the frequency and phase of a second symbol $a_j(t)$, and the mixer output is then averaged over a symbol period using a matched filter or integrator, the output will be zero. This means that with orthogonal signalling it is possible to *increase* the number of symbol states used without affecting the output from an *individual* coherent detector, and hence without increasing the probability of symbol error for each detector.

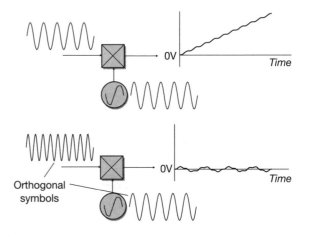

Orthogonal symbols

0V

Time

0V

Time

As we increase the number of orthogonal symbols used for transmission, we can increase the duration of each symbol for a given data information rate (see Section 2.3). The longer the symbol duration, the greater the time for averaging each symbol in the receiver and the better the S/N ratio at the detector output, improving the probability of correct symbol detection. Orthogonal FSK in theory can have any number of orthogonal symbol states, but does so at the expense of ever-increasing occupied bandwidth.

Orthogonal FSK detection

A typical M-ary FSK detector consists of a bank of *correlators* (*mixers with coherent carrier references*), followed by a decision circuit at the output determining which correlator has the largest output and hence which symbol was sent.

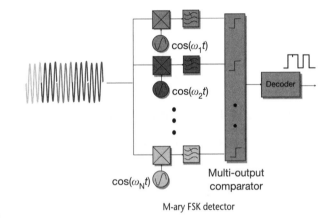

M-ary FSK detector

As the number of symbol states employed tends to infinity, the symbol averaging time becomes very large, reducing the effect of noise to almost zero. The E_b/N_0 required for error-free transmission will thus approach the Shannon–Hartley limit of $-1.6\,\text{dB}$ E_b/N_0 for which error-free communication can be achieved, regardless of how many symbol states and hence how much signalling bandwidth is used.

BER performance for M-ary orthogonal FSK

Shown here are the BER curves for M-ary orthogonal FSK. As predicted, as the number of symbol states is increased, the BER performance improves (at the expense of bandwidth), but never crosses the $-1.6\,\text{dB}$ performance bound.

For digital communications applications where the optimum performance in noise is required, for example in deep space missions where the path loss is so great, M-ary FSK is a very effective modulation technique.

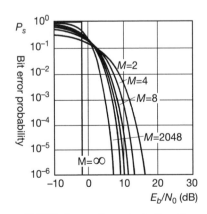

M-FSK(orthogonal): *see in-depth section*

IN DEPTH

Probability of symbol error for M-ary orthogonal FSK

Owing to the non-linear modulation process involved in M-ary FSK, the error probability is difficult to derive and for the general case must be calculated numerically. The general expression as a function of the number of symbol levels M for orthogonal FSK is simply quoted here. Further information can be found in Lindsey and Simon (1972).

$$P_e = 1 - \frac{1}{\pi^{M/2}} \int_{-\infty}^{\infty} e^{-z^2} \left(\int_{\infty}^{z+\sqrt{E_s/N_0}} e^{-y^2} dy \right)^{M-1} dz$$

6.4 M-ary Phase Shift Keying (M-ary PSK)

Quadrature Phase Shift Keying (QPSK)

We have already seen in the case of M-ary FSK that an orthogonal symbol set makes it possible to send two or more symbols simultaneously over a channel, without their affecting the coherent detection performance of any individual symbol. It turns out that as well as the M-ary FSK orthogonal symbol set, there is orthogonality between a *cosine* and *sine* carrier term when averaged over a full number of carrier cycles. This would imply that if we were to signal using binary PSK on the *cosine* of a carrier, and *simultaneously* send a second PSK signal using the *sine* of a carrier, then it would be possible to detect each one independently of the other (as if the other were not there), providing that each detector averaged over a symbol period that contained a whole number of carrier cycles.

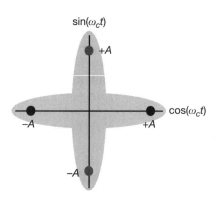

We can thus envisage a PSK modulation scheme with four phase states, 0°, 90°, 180° and 270° which are in phase quadrature with each other. This 4-ary PSK scheme is thus called *Quadrature Phase Shift Keying (QPSK)*. The orthogonality property of QPSK means that it can be used to send information at *twice the speed of BPSK* in the same bandwidth, without compromising the detection performance over BPSK.

Implementation of QPSK – modulator

The block diagram of a QPSK modulator and detector is shown here and is simply two BPSK systems using quadrature carriers summed in parallel. The source data is first split into two data streams (often by allocating alternate bits to the upper and lower modulators), with each data stream running at half the rate of the input data stream. Conventional root raised cosine filtering (see Section 3.4) would be used to shape the data pulses in each channel prior to modulation.

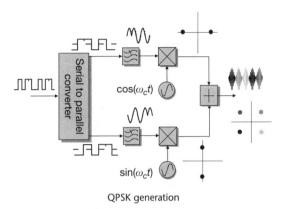

QPSK generation

The constellation diagram for the composite transmitted symbol set shows that the phase states *observed in the channel* would be 45° phase rotated with respect to the individual BPSK sources.

Implementation of QPSK – demodulator

A coherent QPSK receiver requires accurate carrier recovery using a 4th power process, to restore the 90° phase states to modulo 2π.

In addition, an accurate symbol timing recovery circuit (see Section 5.2) is needed for sampling the demodulated filtered data stream. It is possible to use the same methods of symbol timing recovery as proposed for baseband or binary modulation formats.

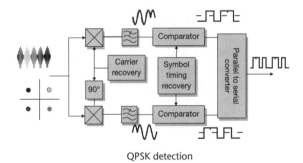

QPSK detection

Data at the output of the comparators is reconstructed into a single data stream using a parallel to serial converter.

Bit error performance of QPSK

As predicted earlier, the *bit* error rate performance for QPSK is theoretically identical to that for BPSK. If the carrier reference is not perfectly phase coherent, however, not only will the wanted signal output voltage of each detector fall, but each detector will suffer *crosstalk* from the orthogonal symbols and the performance will be degraded even further. QPSK thus has lower tolerance to *phase jitter* in the carrier recovery process than BPSK.

The major attraction of QPSK is that it allows the modem designer that opportunity to recover the bandwidth efficiency (see Section 2.2) of a binary *baseband* data stream in a *bandpass* modulation format, *without* sacrificing BER performance.

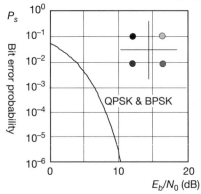

QPSK(bit) = BPSK(bit): $P_s = 0.5erfc[\sqrt{(E_b/N_0)}]$

> *Maximum bandwidth efficiency (QPSK)*
> = 2 bits/second/Hz

Symbol error performance for QPSK

Many textbooks present results for the *symbol* error rate for QPSK which indicate a worse performance than for BPSK. This is to be expected since the symbol states are closer together. It must be remembered, however, that QPSK is conveying *two bits of information for every symbol* and the likelihood of both bits being detected in error is much smaller than only one bit being in error (assuming Gray coding (see Section 3.5) of the symbol states).

Taking this into account, the *bit* error probability will be less than the *symbol* error probability for QPSK, and as we already know is the same as the bit error rate for BPSK.

Because of the good bandwidth efficiency and noise performance, QPSK and its variants are currently the most widely used modulation types in both wired and wireless modems.

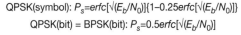

QPSK(symbol): $P_s = erfc[\sqrt{(E_b/N_0)}]\{1 - 0.25erfc[\sqrt{(E_b/N_0)}]\}$
QPSK(bit) = BPSK(bit): $P_s = 0.5erfc[\sqrt{(E_b/N_0)}]$

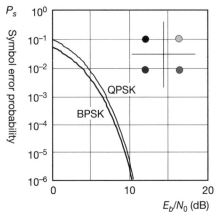

Differential QPSK (DQPSK)

As for BPSK, phase ambiguity (see Section 5.4) in the carrier recovery process for QPSK requires the use of known preambles, differential encoding/decoding (see Section 5.4) of the data, or differential encoding coupled with differential phase detection.

As for BPSK, differential demodulation of QPSK (DQPSK) can be used in preference to coherent detection if simpler implementation is important. The DQPSK modulator uses the same differential data encoder

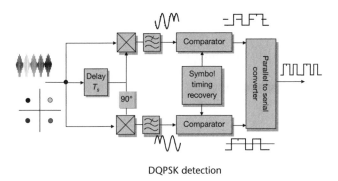

DQPSK detection

for each parallel data stream as the DPSK binary counterpart (see Section 5.4), and likewise employs the same principle of using a 1 symbol delayed version of the received symbol stream to act as the reference for demodulation.

BER performance of DQPSK

The bit error rate performance for DQPSK as a function of E_b/N_0 can be seen to be significantly inferior to that for binary DPSK, and unfortunately does not share the unique property of coherent QPSK in giving no performance penalty for the gain in bandwidth efficiency. This is largely owing to the noise accompanying the delayed demodulator reference mixing with the orthogonal symbols and causing crosstalk in the demodulator. As a result, there is less motivation for the designer to use DQPSK as the effort needed to implement the carrier recovery for coherent detection will yield an approximate 3 dB improvement in performance.

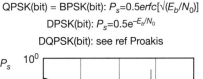

QPSK(bit) = BPSK(bit): $P_s=0.5erfc[\sqrt{(E_b/N_0)}]$

DPSK(bit): $P_s=0.5e^{-E_b/N_0}$

DQPSK(bit): see ref Proakis

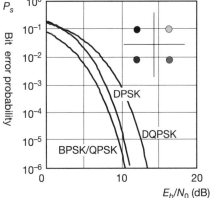

$\pi/4$ QPSK

A recent variant of QPSK, now widely used in the majority of digital radio modems, is the *$\pi/4$ QPSK* format, so called because the four-symbol set is *rotated* by $\pi/4$ or 45° at every new symbol transition. The reason for this rotation is to ensure that the modulation envelope of the filtered QPSK signal never passes through zero, and hence

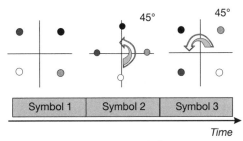

$\pi/4$ rotating symbol set

the peak to mean ratio of the modulation is minimized.

Comparing the vector diagrams for QPSK and $\pi/4$ QPSK, this property is clearly evident. The fact that the modulation envelope does not pass through zero is extremely important for the design of radio power amplifiers. Traditionally, RF power amplifiers are extremely difficult to design with linear response extending down to zero power output, hence the favoured status of $\pi/4$ QPSK over conventional QPSK for radio applications.

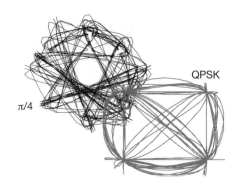

Offset QPSK (OQPSK)

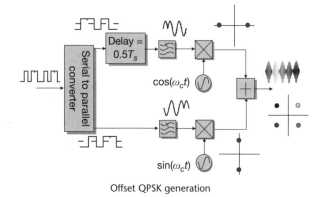

Offset QPSK generation

Offset QPSK can be used in place of $\pi/4$ QPSK to achieve the same end, that is, a non-zero envelope in the modulated signal. Offset QPSK is implemented by staggering the input data streams to the two quadrature BPSK modulators by half a symbol period. The remainder of the modulator and demodulator circuits remains identical to that for conventional QPSK. (The staggered timing must be removed in the receiver.)

For both $\pi/4$ QPSK and Offset QPSK, the BER performance, assuming ideal coherent detection, is identical to that of QPSK itself.

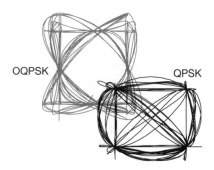

Spectrum of M-ary PSK

Shown here is a comparison of the bandwidth occupancy for *unfiltered* QPSK and Minimum Shift Keying (see Section 5.3). Both schemes have a constant envelope property and thus are tolerant to TX/RX gain distortion. While MSK has a slightly wider main lobe than QPSK, the side-lobe power falls off much more rapidly than that for QPSK, making this system more attractive when adjacent channel interference (see Section 4.3) is an important design consideration.

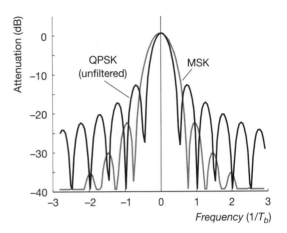

Of course, the QPSK signal can be filtered to reduce the side-lobe energy to any value desired, but will inevitably introduce amplitude variations in the filtered waveform, requiring much more linear processing in the communications link. The data feeding the MSK modulator can also be filtered to reduce the side-lobe energy a little further, an example being GMSK.

Performance of M-ary PSK

Increasing the number of symbol states for M-ary PSK beyond four allows further improvements in bandwidth efficiency, but the additional symbol states are no longer orthogonal (they do not lie on the sine or cosine axis of the constellation diagram). The result is that the performance in noise for $M > 4$ degrades rapidly as M increases.

M-PSK(symbol): P_s (approx.) = $erfc[\sqrt{(\log_2 M . E_b/N_0)} . \sin(\pi/M$

M-PSK(bit) ≈ M-PSK(symbol)/$\log_2 M$

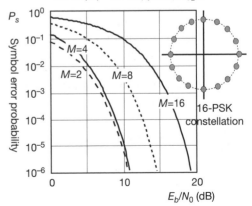

> The bandwidth efficiency for M-ary PSK is:
>
> *Maximum bandwidth efficiency*
> *(M-ary PSK)* = $\log_2 M$ bits/second/Hz

6.5 Combined Amplitude and Phase Keying (QAM/APK)

Introduction

To date, we have considered only *single property* modulators using either phase, amplitude or frequency symbols for conveying the data. One might think that a modulation method combining two or more symbol types could give improved performance in the inevitable trade-off between bandwidth efficiency and noise performance and this is indeed the case.

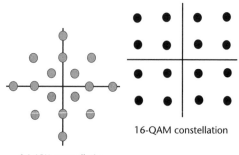

16-APK constellation

16-QAM constellation

The most commonly used combination is amplitude and phase signalling, sometimes classed as *M-ary APK* and sometimes as *Quadrature Amplitude Modulation* (*QAM*), depending on the constraints put on the amplitude/phase relationship.

QAM generation

The simplest form of QAM is in fact the QPSK symbol set, which can be viewed as two quadrature amplitude modulated carriers, with amplitude levels of $+A$ and $-A$. Increasing the number of amplitude levels on each carrier to 4, for example $\pm A$, $\pm 3A$, gives 16 possible combinations of symbols at the transmitter output, each equally spaced on the constellation diagram, and each represented by a unique amplitude and phase.

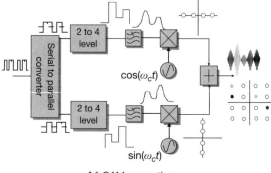

16-QAM generation

Pulse shaping is performed by filtering the multi-level baseband input symbol streams in exactly the same manner as would be used for a binary ASK waveform.

The modulator is again making use of orthogonality of the sine and cosine carriers to allow independent detection of the two M-ary ASK data streams at the receiver.

EXAMPLE 6.1

A digital television system has a source analogue video signal with bandwidth extending from 0 Hz to 2 MHz. This signal is sampled at four times the highest frequency using a 16-bit A/D converter. The resulting data signal is sent over the air using a 16-QAM modulation format with a roll-off factor on the pulse-shaping filters of $\alpha = 0.5$. What is the bandwidth occupied by the transmitted digital video signal?

Solution

The sampling rate for the A/D converter is $4 \times 2\,\text{MHz} = 8$ million samples per second. Each sample is encoded as a 16-bit word, resulting in a bit rate of 16×8 million $= 128\,\text{Mbps}$ at the converter output.

A 16-QAM modulation format conveys 4 bits per symbol, and because it is a bandpass format, this equates to a maximum bandwidth efficiency of 4 bits/second/Hz for an ideal ($\alpha = 0$) filter.

For $\alpha = 0.5$, the bandwidth efficiency is reduced by a factor $(1 + \alpha)$ to 2.66 bits/second/Hz, and the bandwidth required to support a data rate of $128\,\text{Mbps} = 48\,\text{MHz}$.

QAM detection

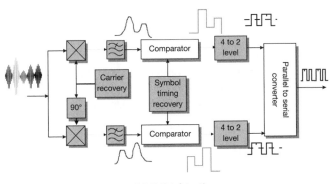

16-QAM detection

QAM can be decoded using *coherent* or *differentially coherent detection* just as for PSK systems. We shall concern ourselves here only with true coherent detection. Details of differential detection systems can be found in Proakis (1989). Just as for QPSK, a QAM demodulator requires the recovery of quadrature carriers in the receiver. Very similar Nth power carrier recovery circuits can be used, although the problem of dealing with phase ambiguity (see Section 5.4) in the recovered carrier is greatly complicated by the presence of the amplitude component of the data symbols.

The output of each demodulator is a baseband multi-level symbol set. Ideally this should undergo matched filtering (see Section 3.5) for optimum performance in noise, before being passed through a bank of comparators to determine the level from each demodulator at the sampling instant, and hence decode the corresponding bit pattern.

M-ary QAM vs M-ary PSK

Comparing the constellation diagrams of M-ary QAM with M-ary PSK we can see that the spacing between symbol states for QAM is greater than that for PSK which is restricted to having symbol states of equal amplitude and thus on a circle equidistant from the origin.

The constellation diagrams shown here have been drawn to scale for *equal average symbol power* for both QAM and PSK systems, and the larger spacing between symbols for QAM means that the detection process should be less susceptible to noise.

The *peak* power for QAM under these conditions is, however, greater than that for M-ary PSK and this must be taken into account if the transmission process is peak power limited.

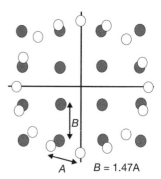

$B = 1.47A$

Comparison of 16-PSK and 16-QAM for equal average symbol power

EXAMPLE 6.2

Draw the constellation diagrams for square 16-QAM.

If the maximum vector length in a square 16-QAM constellation is 100 V rms, determine the long-term average power that would be delivered into a 50-ohm antenna load if each point in the constellation has an equal probability of transmission.

Solution

With reference to one quadrant of the 16-QAM constellation, the average power developed by each of the vectors *A, B, C, D* is as follows:

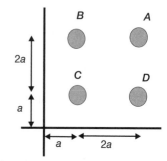

$$A^2 = (3a)^2 + (3a)^2 = 18a^2$$

$$B^2 = D^2 = (3a)^2 + (a)^2 = 10a^2$$

$$C^2 = (a)^2 + (a)^2 = 2a^2$$

$$\text{Average power} = \frac{18a^2 + 2 \times 10a^2 + 2a^2}{4R}$$

The maximum vector length,

$$A = 100\,\text{mV} = \sqrt{18a^2}$$

Therefore

$$a = \sqrt{\frac{(100\,\text{mV})^2}{18}} = 23.6\,\text{mV}$$

Therefore the average power for all symbol states is:

Average power $= 10a^2/R = 111\,\text{W}$

EXAMPLE 6.3

A transmitter for a digital radio system is peak power limited to 150 W. Determine the average power that can be supported for both 16-PSK and square 16-QAM transmission.

Solution

With reference to one quadrant of the 16-QAM constellation, the average power developed by each of the vectors **A, B, C, D** is as follows:

$$A^2 = (3a)^2 + (3a)^2 = 18a^2$$
$$B^2 = D^2 = (3a)^2 + (a)^2 = 10a^2$$
$$C^2 = (a)^2 + (a)^2 = 2a^2$$

$$\text{Average power} = \frac{18a^2 + 2 \times 10a^2 + 2a^2}{4R}$$

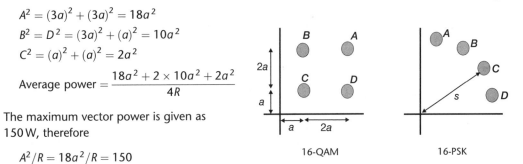

16-QAM

16-PSK

The maximum vector power is given as 150 W, therefore

$$A^2/R = 18a^2/R = 150$$

Therefore

$$a = \sqrt{\frac{150 \times 50}{18}} = 20.4\,\text{W}$$

Therefore the average power for all symbol states is:

Average power$_{\text{QAM}} = 10a^2/R = 83.33\,\text{W}$

The average power for 16-PSK is the same for all symbol states and is equal to the peak symbol power since unfiltered PSK is a constant envelope modulation format. Thus:

Average power$_{\text{PSK}} = s^2/R = 150\,\text{W}$

BER performance for QAM

Comparing the BER curves for PSK and QAM, the intuitive advantage of QAM over PSK drawn from the constellation diagram is borne out, with an approximately 3.5 dB gain in noise immunity for 16-QAM compared with 16-PSK. This improvement comes at the expense of a somewhat more complicated modem design, needing to handle both amplitude and phase information, and to combat both amplitude and phase errors on the channel. In practice, the performance benefits usually outweigh the complexity such that QAM is frequently used in preference to PSK modulation.

If the channel amplitude distortion is particularly severe, such as to dominate errors due to noise, then PSK could prove superior under these conditions.

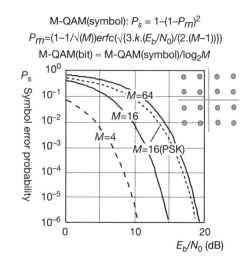

M-QAM(symbol): $P_s = 1-(1-P_m)^2$

$P_m = (1-1/\sqrt{M})erfc(\sqrt{\{3.k.(E_b/N_0)/(2.(M-1))\}})$

M-QAM(bit) $\approx$ M-QAM(symbol)/$\log_2 M$

P_s Symbol error probability

E_b/N_0 (dB)

M-ary Amplitude and Phase Keying (M-ary APK)

In some applications, the constraint of symmetry imposed by QAM systems may not best suit the characteristics of the channel or the detection process. Nonetheless, the designer wishes to have the freedom to place the symbol points anywhere in the constellation diagram, implying both amplitude and phase modulation.

The constellation shown here is that used in the V29.bis telephone modem standard, and correctly belongs to the more general *Amplitude and Phase Keying (M-ary APK)* class rather than the QAM subset class. The reason for this particular symbol layout is to maximize the phase difference between symbols of the *same energy* to 90° rather than only 37° for 16-QAM at the expense of increased amplitude levels. This is on the basis that there is predominantly phase distortion in a telephone line.

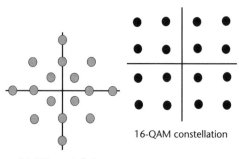

16-APK constellation

16-QAM constellation

Interestingly, the later V32.bis modem standard uses square 16-QAM modulation, but also uses more sophisticated *equalization* (see Chapter 4) and *coding* (see Chapter 7).

Circular QAM constellations are sometimes used in order to alleviate the problems of phase ambiguity in carrier recovery systems and to facilitate differential detection by reducing the number of amplitude levels in the APK format.

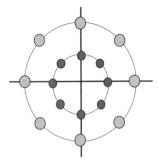

Circular QAM constellation

Gray coding

The concept of Gray coding for optimizing the bit error probability for a given symbol error probability was first introduced in Chapter 3 in the context of M-ary baseband signalling.

For M-ary bandpass signalling it is equally sensible to use an intelligent assignment of bit patterns to symbols in the constellation space to ensure that neighbouring symbols (those most likely to be detected in error) only differ by one bit.

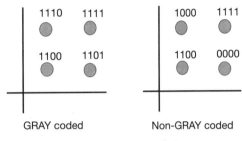

Part of 16-QAM constellation

6.6

Relative performance of multi-level bandpass modulation formats

Symbol error curves for M-ary formats

Shown here are the BER curves for the various families of M-ary systems discussed in this chapter. The QAM formats evidently get us closest to the Shannon capacity limit (see Section 2.4) when we are looking for high spectral efficiency.

Care should be taken when comparing these systems in terms of their bit error probability as this will depend on whether Gray coding has been used and how many errors are restricted to adjacent Gray-coded symbols only. Assuming that only one bit error occurs per symbol, the *bit error rate = symbol error rate/k* (where *k* is the number of bits per symbol).

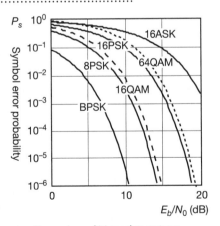

Comparison of M-ary data systems

Further information on the practical design and performance of multi-level modulation formats is available on the Wireless Systems International website (www.wsil.com).

The Shannon limit – how close can we get?

Plotting the performance of M-ary systems on the Shannon diagram demonstrates that none of them is able to deliver the theoretical maximum capacity predicted by the Shannon–Hartley equation (see Section 2.4), being adrift in most cases by 4 dB or more.

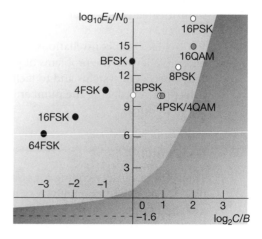

In order to get closer to the Shannon limit we need to introduce 'coding' of the data, allowing us to detect and correct data errors. A simple introduction to coding and its benefits is given in the next chapter.

It is clear that M-ary FSK allows us to exchange bandwidth efficiency to improve the power efficiency of the modem, but the uncoded performance falls very short of the ideal -1.6 dB E_b/N_0 performance limit.

Table of CCITT telephone modem characteristics

In Chapter 1, the influence of standards on digital communications systems design was discussed, and nowhere are standards more rigorously applied than for telephone dial-up modem systems.

In order to ensure interoperability of telephone modems, the International Telecommunications Union (ITU), formerly known as the International Telegraph and Telephone Consultative Committee (CCITT), oversees the standardization of modulation formats, coding

CCITT(ITU) Telephone Standards			
Version no	Bit rate	Modulation format	Protocol
Bell 103	0–300	FSK	async
Bell 202	1200	FSK	async
V.22	1200/600	QPSK/FSK	async/sync
V.26bis	2400	QPSK	sync
V.27	4800/2400	8-DQPSK/QDPSK	sync
V.29	9600	16-APK	sync
V.32	9600	32-QAM/16-QAM	sync
V.33	14 400	32-QAM	sync
V.34	33 600	>1024-QAM	sync
V.90	56 000	>1024-QAM	sync

schemes and so on for all telephone commercial modems. The table shown here lists most of the standard types, showing how the complexity and number of symbol states increase as the throughput increases.

This table also provides an interesting insight into the development of modem technology. The early low speed modems used the very simple, but bandwidth and power inefficient, non-coherent FSK modulation format, while the most recent modems use the constellation space to the full and employ some of the most powerful DSP processors in the market-place to realize sophisticated equalization, convolutional coding, and so on.

QUESTIONS

6.1 An 8-ary ASK scheme makes use of a root raised cosine filter in both transmitter and receiver, with an α of 0.33. What is the bandwidth required to support a data rate of 64 kbps?

6.2 A digital voice link requires a bit error performance of no worse than 1 error in 10^3. From the BER plots for M-ary ASK, what is the approximate E_b/N_0 value required for a binary ASK and a 16-ary ASK modem?

6.3 A 32-ary ASK modem has a symbol error rate of 2 in 10^5 under worst-case conditions. What is the approximate bit error rate assuming Gray coding has been used?

6.4 A 4-ary orthogonal FSK modem has a symbol rate of 2400 symbols per second. If the lowest symbol frequency is 8 kHz, what will be the other three symbol frequencies?

6.5 What is the maximum bandwidth efficiency possible for a modem required to operate at an E_b/N_0 value of -1.2 dB?

6.6 If the peak symbol power for a square 16-QAM system is 200 W, measured in a 50-ohm load, what are the amplitudes of the different symbol vectors in the transmitted waveform? (Neglect any filtering effects.)

6.7 A 64-QAM data link operates at 256 kbps. What is the underlying symbol rate on the channel, and what is the occupied bandwidth if two root raised cosine filters are employed, one in the transmitter and one in the receiver, each with an $\alpha = 0.5$?

6.8 What is the minimum bandwidth required to support a 256 kbps data stream using:

(a) four-level bipolar baseband signalling?

(b) four-level polar baseband signalling?

(c) BPSK?

(d) QPSK?

(e) 64-QAM?

6.9   A microwave line-of-sight communication link uses 256-QAM to convey 32 Mbps. The bandwidth occupied by the signal is 7 MHz.

   (a) What is the value of α used in the raised cosine filtering?

   (b) If the signal to noise ratio on the link is 40 dB, what is the theoretical maximum capacity for the channel in a 7 MHz bandwidth?

6.10 A customer requires a microwave radio link to provide a bit rate of 2 Mbps in a bandwidth of 400 kHz. The minimum signal to noise ratio on the channel is 30 dB.

   Can the channel support the required capacity, and how many symbol states would be required?

7 Coding theory and practice

To fully grasp the subject of coding for digital communications requires a very good working knowledge of mathematics and it is difficult to give an introduction to coding without either quickly losing the reader in complex equations and statistical theory, or providing a superficial overview which lacks rigour. This chapter attempts to inform the reader of the terminology and importance of coding for communications, but goes no further. Good texts on the subject include, of course, Proakis (1989) and Halsall (1992) from different perspectives.

The term *coding* is applied to many operations within a communications system, including:

- *Source coding* – where an analogue or digital source is altered in some way to make it best suited for transmission purposes.

- *Channel coding* – where extra information (redundancy) is added to an existing bit or symbol set in order to provide a means of detecting and/or correcting transmission errors. Usually channel coding involves operations on binary data.

- *Modulation coding* – where a modulation symbol set (constellation) is expanded, again in order to allow the detection and correction of erroneous symbols. Modulation coding usually involves operations on analogue data symbols.

In many modern communication links, combinations of source, channel and modulation coding are employed in a dependent manner to optimize performance.

7.1 Source coding

Introduction

The purpose of source coding is to transform the information type in the source to a form best suited to the transmission process. Often this involves converting an analogue signal such as voice or light intensity in an image to a digital binary representation for transmission using a modem.

These days, the source coding process will usually implement an algorithm to realize bit or symbol content compression in addition to the standard process of quantization and A/D conversion. Standard image compression algorithms are the MPEG and JPEG formats for moving and static images respectively. At the present time, the source coding for music and speech is not so well standardized, with many different formats in use throughout the world.

In most, but not all cases, a complementary decoder is implemented to restore the signal to near or exactly its original format.

Waveform coding – A/D converters

The term *waveform coding* is applied to source encoding methods that seek to digitize the incoming analogue waveform, without making use of any information about its frequency content or source parameters, and thus achieve more efficient coding.

This method is the most flexible source coding technique, able to accommodate any type of input waveform regardless of the generating source, and is typified by the *analogue to digital converter*.

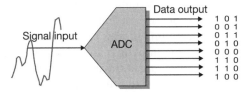

Analogue to digital converter

The A/D conversion process, often more grandly called *pulse code modulation* in many textbooks, usually involves regular sampling of the input signal level and then a conversion of this sampled value into a number representing the level. The accuracy of the representation is governed primarily by the resolution of the A/D converter – that is, how many data bits it uses to represent each measured value. Typical A/D converters use eight bits for telephony voice digitization, and 16 or 18 bits for hifi music digitization. Some professional mixing desks use 24-bit converters! Most communications links will make use of the complementary *digital to analogue (D/A) converter* in order to restore the analogue waveform samples from the received data words.

Nyquist sampling

One of the key goals in waveform sampling is to digitize only the minimum number of samples necessary to represent the waveform accurately and hence allow accurate reconstruction on reception.

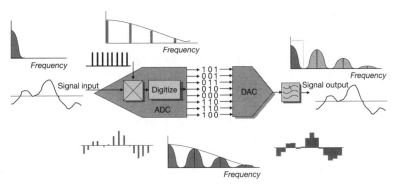

Nyquist sampling system

The minimum rate at which an arbitrary waveform can be sampled *without loss of information* is in fact twice the bandwidth of the input waveform.

This is known as the *Nyquist sampling criterion*. Sampling at less than twice the bandwidth of the input signal (equivalent to twice the maximum modulation frequency for baseband signals) results in what is termed *aliasing*.

The Nyquist sampling requirement can be derived intuitively from our knowledge of Fourier series summarized in Chapter 1. We can view the sampling process as the mixing of the input signal with a train of very narrow data sampling pulses which will result in sum and difference components appearing at the mixer output for each harmonic of the pulse waveform mixing with the signal waveform as shown. This is the spectrum that would effectively appear at the output of a D/A converter. In order to reconstruct the input waveform correctly, the D/A output needs to be filtered so that only the spectral components present within the source signal remain.

Nyquist sampling theorem – aliasing

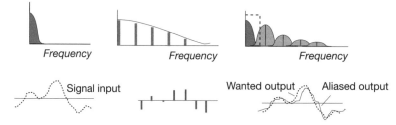

Example of aliasing in a sampled system

If the Nyquist sampling criterion is not met, the effect is for the sum and difference components associated with each harmonic of the sampling signal to overlap with those of adjacent harmonics in the sampling spectrum. Clearly it is now not possible to filter out the wanted from the unwanted signals and thus perfect reconstruction of the original signal is not achieved.

It is immediately apparent from this analysis that sampling at twice the maximum input signal frequency is thus the minimum sampling rate needed for A/D conversion. It is also evident that removal of the possible alias components when sampling at this minimum rate requires a 'brick wall' filter for signal recovery after D/A conversion. In practice, a sampling rate of at least three times the maximum sampling frequency is used in order to reduce the specification of these 'anti-aliasing' filters.

Some modern A/D and D/A converters, called sigma–delta converters, use many-fold oversampling ($\times 64$ or $\times 128$ is typical) with in-built decimation or interpolation filters.

E | **EXAMPLE 7.1**

A broadcast audio source signal contains frequencies in the range from 50 Hz to 18 kHz. What is the minimum sampling rate required for an A/D converter in order to ensure that there will be no aliasing? What is a practical sampling rate to choose for this application?

If by accident a high frequency tone at 30 kHz is added into the audio source, at what frequency will this signal appear in the sampled waveform if the sampling rate is set at 40 000 samples per second?

Solution

The Nyquist sampling rate for perfect signal reconstruction (no aliasing) must be twice the highest frequency component of a baseband signal, that is, 2×18 kHz $= 36\,000$ samples per second. This sampling rate assumes that a 'brick wall' low pass filter can be used to remove alias components. In practice, a sampling rate of 44 100 samples per second is commonly used in the hifi industry as the standard sampling rate for high quality audio signals with frequencies up to 20 kHz.

If a sampling rate of 40 000 is used to digitize a tone with a frequency of 30 kHz, the sampling process will produce a difference frequency component at $(40\,000 - 30\,000) = 10\,000$ Hz, which falls well within the half Nyquist bandwidth of 20 kHz for a 40 kHz sampled system. It would thus appear that a 10 kHz tone had been applied at the input to the A/D converter.

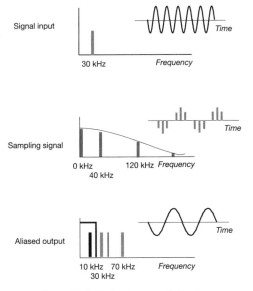

Example of aliasing in a sampled system

Dynamic range

The ability of an A/D converter to cope with both large and small signals is an important factor in waveform encoding, and the ratio of V_{max} to V_{min} over which a converter will operate is called the *dynamic range*. This parameter depends heavily on the resolution of the A/D, that is, the number of bits available to represent any given sample. The more bits in the converter, the more *quantization* levels the converter is using to match to any given waveform sample.

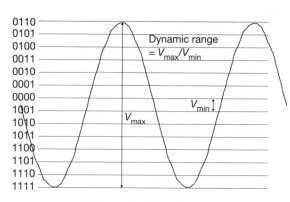

A/D converter dynamic range

It is not difficult to see that an *n*-bit converter can differentiate between $2^n = M$ discrete signal levels and that the minimum signal variation that it can detect and represent is V_{max}/M volts. This is often called the *quantization step size*.

The dynamic range for an A/D converter is thus given by:

$$Dynamic\ range = V_{max}/V_{min} = V_{max}/(V_{max}/M) = M\ or\ 2^n$$

Expressed in dB, this gives us the well-known formula: the dynamic range of a linear A/D is $n \times 6.02\,dB$.

An 8-bit converter thus has a dynamic range of approximately 48 dB.

Quantization noise

Another very important parameter in any source encoding scheme is the level of noise and distortion introduced by the coding process. For waveform encoding, the main noise source is quantization error, that is, the amplitude errors which the A/D and D/A conversion process introduces into the signal by not having infinite precision. The level of quantization noise is dependent on how close any particular sample is to one of the *M* levels in the converter. For a

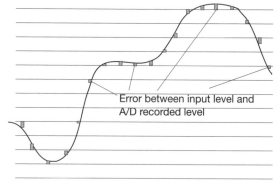

Error between input level and A/D recorded level

Quantization error in A/D converters

speech input, this quantization error will be manifest as a noise-like disturbance at the output of a D/A converter.

> The *signal to quantization noise ratio* for an A/D converter is thus clearly a function of the number of bits used and is given by (see in-depth section):
>
> *Peak S/N ratio* $= 3M^2/2$

An 8-bit converter thus has a signal to quantization noise ratio of approximately 50 dB for a full-scale input signal.

IN DEPTH

Calculation of signal to quantization noise ratio

The maximum quantization error q that can occur in the sampled output of an A/D converter is equal to half the minimum resolution of the converter which is given by $q = V_{max}/M$, where $M = 2^n$ is the number of quantization levels for the converter.

Making the assumption that all values of quantization error within the range $+q/2$ to $-q/2$ are equally likely, the mean-squared value of the quantization error, E_q, can be calculated from,

$$E_q = \frac{1}{q} \int_{-q/2}^{q/2} (error)^2 \cdot d \cdot error = \frac{q^2}{12}$$

The rms error is thus $q/\sqrt{12}$.

The peak signal voltage for a sinewave input will be $V_{max} = M \cdot q/2$ (allowing for positive and negative excursions of the sinewave), and the rms value is thus $M \cdot q/2\sqrt{2}$. The signal to quantization noise power ratio can now be determined as:

$$S/N_q = \frac{(Mq/2\sqrt{2})}{(q/\sqrt{12})^2} = \frac{3M^2}{2}$$

Thus, for an 8-bit converter, the peak signal to quantization noise ratio is approximately $3 \times (2^8)^2 \times 0.5 = 98.304$ or 50 dB.

Companding

A method of reducing the number of bits required in a converter while achieving an equivalent dynamic range or signal to quantization noise ratio is to use a technique known as *companding*. The term 'companding' comes from a combination of the words COMPressing and expANDING, which adequately describe the process involved.

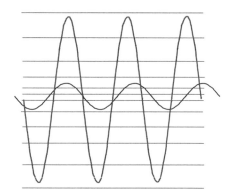

Companding of the quantization levels in A/D converters

Essentially, in order to improve the resolution of weak signals within a converter, and hence enhance the signal to quantization noise ratio, the weak signals need to be enlarged, or the quantization step size decreased, but only for the weak signals. Strong signals, on the other hand, can potentially be reduced without significantly degrading the signal to quantization noise ratio, or alternatively the quantization step size is increased. This compression process must be matched with an equivalent expansion process in the D/A converter if the waveform integrity is to be maintained.

Because this technique is so effective at reducing the number of A/D and D/A bits needed to provide adequate signal to quantization noise ratio for speech signals in particular, international standards have been set defining the compression and expansion ratios to be used for telephone interconnect throughout the world. There are in fact two standards, one predominantly a US standard called μ-law companding, and the other a European and ITU standard called *A-law* companding.

EXAMPLE 7.2

A telephone system uses filters to band limit the signal from each user to a maximum of 4 kHz. The maximum signal generated by each user is 1 V, and the minimum signal that must be supported is 72 dB below this value. What is the minimum bit rate that must be supported by the telephone system for each user on the network?

If companding is used within the A/D conversion process to achieve equivalent dynamic range, but with only eight bits, what will be the bandwidth required of a binary baseband data link for a single digitized voice channel? If 32 channels are multiplexed into an E1 frame, what is the number of bits per frame and the data rate on the E1 link?

Solution

A 4 kHz maximum input signal requires a minimum sampling rate of 8000 samples per second. In order to maintain a 72 dB dynamic range, the A/D converters must use $72/6 = 12$ bits, giving a total bit rate per telephone channel of $8000 \times 12 = 96\,000$ bps.

With the use of companding, the bit rate per channel reduces to 8×8000 = 64 kbps. For a baseband channel, the maximum bandwidth efficiency is 2 bits/second/Hz for binary signalling, hence 32 kHz of bandwidth is needed per voice band.

For 32 channels per frame, the frame length of the E1 line is $32 \times 8 = 256$ bits. The frame rate must equal the sampling rate of any one of the channels, that is, 8000 frames per second, hence the E1 data rate = $256 \times 8000 = 2048$ Mbps.

A description of the E1 and T1 channel types is given in Section 1.3.

Voice coding

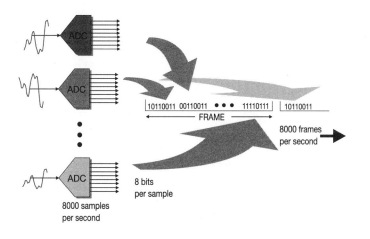

The most widely used form of source encoding to date is in fact based on these μ-law and A-law companding A/D and D/A converters, often referred to as *codecs*, which form the digital interface for almost every telephone line in the world. At the telephone exchange or switching centre, each analogue signal from the domestic phone is converted using an 8-bit μ-law or A-law codec, with a standardized sampling rate of 8000 times per second. (The maximum voice frequency is limited to 3400 Hz, hence the Nyquist criterion is met.) This results in a data rate of 64 kbps for each voice link.

At the exchange, a number of these eight-bit data words from different phone sources are collected (multiplexed) into a 'frame' (32 for E-type systems and 24 for A-type systems), and then sent using either baseband signalling or bandpass signalling methods (see Chapters 3 and 6) over the national and international exchange links. In order to keep pace with the codec sampling rate, a new frame must be constructed and sent every 1/8000 seconds.

IN DEPTH

Typical Codec IC for telephone signal digitization. Data sheet courtesy of Motorola.

MOTOROLA
SEMICONDUCTOR TECHNICAL DATA

Order this document
by MC145554/D

PCM Codec-Filter

The MC145554, MC145557, MC145564, and MC145567 are all per channel PCM Codec–Filters. These devices perform the voice digitization and reconstruction as well as the band limiting and smoothing required for PCM systems. They are designed to operate in both synchronous and asynchronous applications and contain an on–chip precision voltage reference. The MC145554 (Mu–Law) and MC145557 (A–Law) are general purpose devices that are offered in 16–pin packages. The MC145564 (Mu–Law) and MC145567 (A–Law), offered in 20–pin packages, add the capability of analog loopback and push–pull power amplifiers with adjustable gain.

These devices have an input operational amplifier whose output is the input to the encoder section. The encoder section immediately low–pass filters the analog signal with an active R–C filter to eliminate very–high–frequency noise from being modulated down to the pass band by the switched capacitor filter. From the active R–C filter, the analog signal is converted to a differential signal. From this point, all analog signal processing is done differentially. This allows processing of an analog signal that is twice the amplitude allowed by a single–ended design, which reduces the significance of noise to both the inverted and non–inverted signal paths. Another advantage of this differential design is that noise injected via the power supplies is a common–mode signal that is cancelled when the inverted and non–inverted signals are recombined. This dramatically improves the power supply rejection ratio.

After the differential converter, a differential switched capacitor filter band passes the analog signal from 200 Hz to 3400 Hz before the signal is digitized by the differential compressing A/D converter.

The decoder accepts PCM data and expands it using a differential D/A converter. The output of the D/A is low–pass filtered at 3400 Hz and sinX/X compensated by a differential switched capacitor filter. The signal is then filtered by an active R–C filter to eliminate the out–of–band energy of the switched capacitor filter.

These PCM Codec–Filters accept both long–frame and short–frame industry standard clock formats. They also maintain compatibility with Motorola's family of TSACs and MC3419/MC34120 SLIC products.

The MC145554/57/64/67 family of PCM Codec–Filters utilizes CMOS due to its reliable low–power performance and proven capability for complex analog/digital VLSI functions.

MC145554/57 (16–Pin Package)
- Fully Differential Analog Circuit Design for Lowest Noise
- Performance Specified for Extended Temperature Range of – 40 to + 85°C
- Transmit Band–Pass and Receive Low–Pass Filters On–Chip
- Active R–C Pre–Filtering and Post–Filtering
- Mu–Law Companding MC145554
- A–Law Companding MC145557
- On–Chip Precision Voltage Reference (2.5 V)
- Typical Power Dissipation of 40 mW, Power Down of 1.0 mW at ± 5 V

MC145564/67 (20–Pin Package) — All of the Features of the MC145554/57 Plus:
- Mu–Law Companding MC145564
- A–Law Companding MC145567
- Push–Pull Power Drivers with External Gain Adjust
- Analog Loopback

MC145554
MC145557
MC145564
MC145567

16   1   **L SUFFIX**
CERAMIC PACKAGE
CASE 620
MC145554/57

16   1   **P SUFFIX**
PLASTIC DIP
CASE 648
MC145554/57

16   1   **DW SUFFIX**
SOG PACKAGE
CASE 751G
MC145554/57

20   1   **L SUFFIX**
CERAMIC PACKAGE
CASE 732
MC145564/67

20   1   **P SUFFIX**
PLASTIC DIP
CASE 738
MC145564/67

20   1   **DW SUFFIX**
SOG PACKAGE
CASE 751D
MC145564/67

REV 1
9/95 (Replaces ADI1517)

M MOTOROLA

PIN ASSIGNMENTS

MC145554, MC145557

VBB	1●	16	VFXI+
GNDA	2	15	VFXI−
VFRO	3	14	GSX
VCC	4	13	TSX
FSR	5	12	FSX
DR	6	11	DX
BCLKR/CLKSEL	7	10	BCLKX
MCLKR/PDN	8	9	MCLKX

MC145564, MC145567

VPO+	1●	20	VBB
GNDA	2	19	VFXI+
VPO−	3	18	VFXI−
VPI	4	17	GSX
VFRO	5	16	ANLB
VCC	6	15	TSX
FSR	7	14	FSX
DR	8	13	DX
BCLKR/CLKSEL	9	12	BCLKX
MCLKR/PDN	10	11	MCLKX

FUNCTIONAL BLOCK DIAGRAM

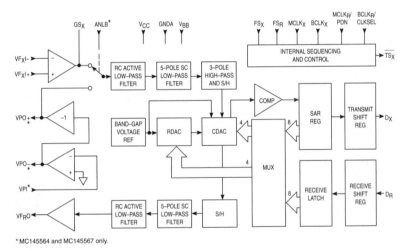

* MC145564 and MC145567 only.

The source coding discussion so far has focused on a particular type of linear waveform encoding which, while very flexible and widely used, leads to a high data transmission rate (64 kbps per telephone voice channel) for the encoded signal. In many applications where transmission capacity is limited,

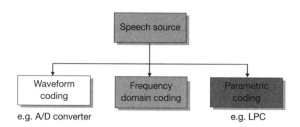

often because of restricted channel bandwidth, it is desirable to achieve a much lower data rate for the encoded signal, hopefully without incurring any significant degradation in the perceived (subjective) quality of the signal.

If we are not concerned with preserving the waveform shape perfectly, but rather we wish to maintain the subjective quality of the received signal – visual or audio fidelity – then we can move away from waveform coding and use more sophisticated *frequency domain* or *source modelling* (parametric) based methods.

The subject of voice coding is vast, and falls outside the scope of this book. A good reference is Papamichalis (1987). A technique that is proving to give good voice quality and good compression performance is based on *Linear Predictive Coding* which in part attempts to model the actual human voice synthesis process. Data rates of the order of 7000 bps are now considered to give acceptable voice quality telephone communications (compared with 64 kbps for the standard waveform codec), and these techniques are used extensively in modern digital cellular systems.

Intelligent source coding

Where the error characteristics of the communications channel are known, it is possible to bias the source encoding/ decoding process deliberately to be tolerant of a certain type of error.

The images shown here are good examples of how intelligent source coding can pay off. The first image uses conventional image coding techniques designed for data channels that exhibit a random distribution of bit errors with time. When subject to a channel that produces short, focused bursts of errors, the decoded image is severely degraded. The second image has been passed through an encoder/decoder designed to be resilient to burst errors, and clearly demonstrates the marked improvement that can be achieved if the source coding is tailored to the channel error conditions – where known.

(Images kindly provided by Professor David Bull – Centre for Communications Research, University of Bristol.)

7.2 Channel coding

Introduction

Channel coding is most often applied to communications links in order to improve the reliability of the information being transferred. By adding additional bits to the transmitted data stream (which of course increases the amount of data to be sent), it is possible to detect and even correct for errors in the receiver.

- *Error detection* – In its most elementary form this involves recognizing that part of the received information is in error and if appropriate or permissible requesting a repeat transmission – ARQ (*Automatic Repeat Request Systems*).

- *Error detection and correction* – With added complexity, it is possible not only to detect errors, but also to build in the ability to correct errors without recourse to retransmission. This is particularly useful where there is no feedback path to the transmission source with which to request a resend. This process is known as *FEC* (*Forward Error Correction*).

Types of ARQ operation

- *Stop and Wait ARQ* – This is the simplest ARQ method where the transmitter waits after *each* message for an acknowledgement of correct reception (known as an ACK) from the receiver. If the message is received in error, a negative acknowledgment (NAK) is returned. While this process is taking place, any new messages must be stored in a buffer at the transmitter site.

- *Go Back N ARQ* – As the name suggests, the transmitter in this case continues to transmit messages in sequence *until* a NAK is received. The NAK identifies which message was in error and the transmitter then 'back-tracks' to this message, starting to retransmit all messages in the sequence from when the error occurred. Clearly, this has less signalling overhead (no ACKs used) than the Stop and Wait protocol.

- *Selective ARQ* – By making the protocol slightly more complex, and by providing a buffer in the receiver as well as the transmitter, it is of course possible for the receiver to inform the transmitter of the specific message or packet that is in error. The transmitter then need only send this specific message which the receiver can reinsert in the correct place in the receiver buffer. Although the most complex, this is also the most efficient type of ARQ protocol and the most widely used. There are several variants of this protocol optimized for a given set of channel characteristics.

Parity

A basic requirement of the ARQ system is for the receiving equipment to be able to detect the presence of errors in the received data. One of the simplest yet most frequently used techniques for detecting errors is the *parity check bit*. Most people who have set up a modem for a computer link will have found a setting for *odd* or *even parity* alongside the number of 'stop bits' and so on.

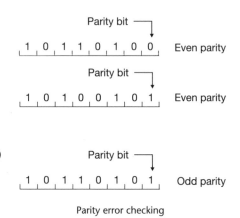

The parity check bit is usually a single bit (1 or 0) appended to the end of a data word such that the number of 1s in the new data word is *even* for even parity, or *odd* for odd parity.

Thus for the first example data word shown here, a 0 must be added as the parity bit for an even parity design because the number of 1s in the word is already even. For the second word, there is an odd number of 1s so a logic 1 parity bit is added to make this number even.

On reception, each data word, with appended parity bit, is checked to see how many 1s are present. For an even parity design, the number must be even. If it is found to be odd, it can be concluded that at least one error has occurred during transmission and the ARQ process can begin. Of course, if two bits are in error, the parity check will pass, and the errors will go undetected.

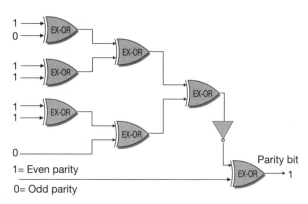

Parity bit generation circuit

The single bit parity check is thus best suited to low noise, low distortion links where the error rate is known to be very low. For links with a high probability of error, more sophisticated error checking methods must be used – the block or convolutional codes described next, requiring the addition of larger numbers of redundant bits.

Shown here is a simple circuit for working out the necessary parity bit for even or odd parity and a seven-bit input data word.

Types of FEC coding

There are two main types of forward error correction coding scheme:

- *Block coding* – where a group (block) of bits is processed as a whole in order to generate a new (longer) coded block for transmission. A complementary block decoder is used in the receiver. Block coding is described in detail in below.

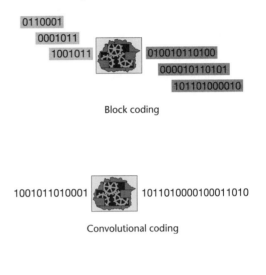

Block coding

- *Convolutional coding* – which operates on the incoming serial bit stream generating a real-time encoded serial output stream. A complementary serial (convolutional) decoder is used in the receiver. Convolutional coding is described in detail in Section 7.5.

1001011010001 [image] 1011010000100011010

Convolutional coding

7.3 Block coding

Basics of block coding

The terminology for block coding is that an input block of k bits gives rise to an output block of n bits and this is called an (n, k) *code*. This increase in block length means that the useful data rate (the information transfer rate) is reduced by a factor k/n. This is called the *rate of the code*.

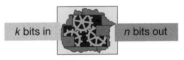

Code rate $R = k/n$

The additional data bits are carefully chosen such that they help differentiate one pattern of k bits in a block from a different pattern of k input bits. The factor $1 - k/n$ is usually termed the *redundancy* of the block code.

Block coding analogy

As an analogy, consider the input blocks of data as being represented by different-sized balls. After encoding, the different-sized balls have been given different colours. The decoder is now optimized for checking *both* size and

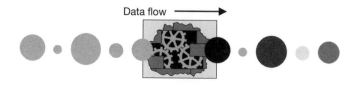

colour. If the receiver detects a particular-sized ball, with the wrong colour, it can now detect the error. It may also be possible to *correct* the error from its stored knowledge of all valid ball sizes and colours.

If the channel errors are so numerous that the ball size and colour are transformed into another valid ball size and colour then the error will go undetected and uncorrected.

Coding efficiency

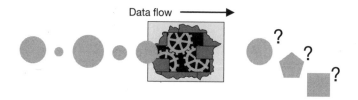

The *efficiency* of a code is a measure of how well errors can be detected or corrected versus the bit overhead required to implement the code. In the ball analogy, colour coding the balls may be very effective, but adding colour may introduce a very high overhead of data, and perhaps changing the ball shape may provide the same error detection or correction capability, but use fewer additional data bits. This would therefore be classed as a more efficient coding scheme.

Certain code types are better at *detecting* errors than *correcting* them, and these are thus well suited to ARQ schemes where we need to know an error has occurred, and our corrective action is to request a retransmision. Equally, there are codes that are best suited for correcting errors and these would be used where no retransmission was possible. A good example for this type of code would be paging devices, or missile control systems!

Hamming codes – an example of a block code

Hamming codes, named after their discoverer, R.W. Hamming, are a well-known type of block code.

A rate $R = 4/7$ Hamming code is illustrated here, with each of the 16 possible four-bit input blocks, coded as seven-bit output blocks. This set of 16 output blocks has been selected from the $2^7 = 128$ possible seven-bit patterns as being most dissimilar. In this case, each output block can be seen to differ from all the other blocks by at least three bits. Hence, if one or two errors occur in the transmission of a block, the decoder will realize that this is not a valid block and flag an error. In the case of only a single bit error occurring, it is also possible for the receiver to match the received block to the closest valid block and thereby *correct* the single error. If three errors occur in a block, the original block may be transformed into a new valid block and all the errors go undetected.

Hamming $R = 4/7$ code		
Block number	Input block data	Output block data
0	0000	000 + 0000
1	1000	110 + 1000
2	0100	011 + 0100
3	1100	101 + 1100
4	0010	111 + 0010
5	1010	001 + 1010
6	0110	100 + 0110
7	1110	010 + 1110
8	0001	101 + 0001
9	1001	011 + 1001
10	0101	110 + 0101
11	1101	000 + 1101
12	0011	010 + 0011
13	1011	100 + 1011
14	0111	001 + 0111
15	1111	111 + 1111

Hamming distance

The number of bits difference between pairs of coded blocks (code words) is a very important property of the code, and is known as the *Hamming distance*. The greater the Hamming distance, the more dissimilar the code words or blocks and the better the chance of detecting or correcting errors.

> A block code with a Hamming distance of p can detect up to $p - 1$ errors, and correct $(p - 1)/2$ errors.

Shown here is a subset of code words from a $(4, 7)$ code which can be seen to have a Hamming distance of 3 between each code word. If this were the minimum distance between all code words in the set then this code would be classed as having a *minimum Hamming distance* $d_{min} = 3$. The code could thus detect up to two errors and correct one error.

4	0010	111 + 0010
5	1010	001 + 1010
6	0110	100 + 0110
7	1110	010 + 1110
8	0001	101 + 0001

BER performance for a (7, 4) Hamming code

The BER curves for uncoded and coded BPSK are shown here for the $k = 4$, $n = 7$ Hamming code. The curves have been normalized for equal energy per information bit (pre-coding), bearing in mind that the energy per transmitted bit is less than the energy per information bit by a factor equal to the *code rate R*. The mathematics required to work out the performance of some of these codes is very complex and often it is only possible to derive approximate results or run simulations over very long periods of time.

The improvement in E_b/N_0 performance of the coded vs uncoded systems, at a specified BER, is termed the *coding gain*.

Note that:

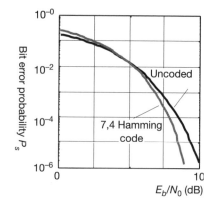

$$E_b(\text{precoded}) = E_b(\text{postcoded})/R$$

Higher-order Hamming codes

A complete family of Hamming codes exist, some of which are shown here. As the length increases, the codes give similar correction performance at a given BER, but with ever-reducing coding overhead. In addition, the asymptotic coding gain increases with increasing code length.

For example, the (4, 7) Hamming code has an asymptotic coding gain of about 0.5 dB while the (120, 127) Hamming code has a coding gain in the order of 1.5 dB. The code rate is higher for the (120, 127) code, $R = 120/127$, than the (4, 7) code, $R = 4/7$, and so has less redundancy.

k	n	$R = k/n$
4	7	0.57
11	15	0.73
26	31	0.84
57	63	0.91
120	127	0.94
247	255	0.968
502	511	0.982

Implementation complexity

As the length of a block code increases, two problems begin to emerge:

• The time taken to collect k bits to form a block increases with increasing block length, introducing delay into the transmission process which may be unacceptable for real-time applications such as voice transmission.

- The decoder complexity increases almost exponentially with block length as the decoder searches through 2^k valid code words to find the best match with the incoming 2^n possible coded blocks. In addition to the complexity, the decoding delay can be significant.

7.4 Advanced block coding

Block code families

Hamming codes are in fact a subset of a more general code family called *BCH* (*Bose–Chaudhuri–Hocquenghem*) codes discovered in 1959 and 1960.

Whereas the Hamming codes can *only* detect up to two errors *or* correct one, the general BCH code family can detect and correct *any number of errors* if the code word used is long enough.

For example, the Hamming $(4, 7)$ code corrects only one error, while the BCH $(64, 127)$ code corrects 10 errors. For real error correcting power, the $(11, 1023)$ code

k	n	Code rate $R = k/n$	No. of bits corrected
4	7	0.57	1
5	15	0.33	3
24	63	0.38	7
64	127	0.5	10
247	255	0.97	1
171	255	0.67	11
11	1023	0.01	255

can correct a staggering 255 errors but with a very high coding overhead indeed. This would be used where reliability of transmission is key and data throughput is less important.

Interleaving

The block codes described thus far work best when errors are distributed evenly and randomly between incoming blocks. This is usually the case for channels corrupted primarily by AWGN, such as a land-line telephone link.

In a mobile radio environment, however, errors often occur in bursts as the received signal fades in and out due to the multipath propagation and the user's motion. In order to distribute these errors more evenly between coded blocks, a process known as *interleaving* is used.

Implementation

One way to accomplish interleaving is to read the encoded data blocks as rows into a matrix. Once the matrix is full (incurring a time delay penalty), the data can be read out in columns, redistributing the data for transmission.

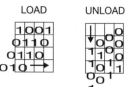

Interleaving Matrix

At the receiver, a *de-interleaving* process is performed using a similar matrix filling and emptying process, reconstituting the original blocks. At the same time the burst errors are uniformly redistributed across the blocks.

The number of rows or columns in the matrix is sometimes referred to as the *interleaving depth*. The greater the interleaving depth, the greater resistance to long fades, but also the greater the *latency* in the decoding process as both the TX and RX matrix must be full before encoding or decoding can occur.

E EXAMPLE 7.3

A mobile radio data link uses interleaving to spread the data errors on reception. If the interleaving depth used is a 10 × 8 matrix, and the bit rate for the signal is 9600 bps, what is the latency introduced by the interleaving process?

Solution

The de-interleaving process requires that the matrix rows are full up before the data can be read out from the columns, thus 8 × 10 bits = 80 bits must be read into the matrix before the de-interleaved data can be extracted. The data rate is 9600 bps, thus the time taken to load the matrix is 80/9600 = 8.3 ms. A further 8.3 ms will be used loading the interleaving matrix in the transmitter, giving a total latency of 16.6 ms.

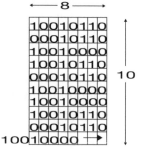

10 x 8 interleaving

Reed–Solomon (RS) codes

RS codes are a subset of BCH codes that operate at the block level rather than the bit level. In other words, the incoming data stream is first packaged into small blocks, and these blocks are then treated as a new set of *k* symbols to be packaged into a *super-coded* block of *n* symbols. The result is that the decoder is

| 1100110 | 1011001 | 0011111 | 1011110 | 0111000 | 0000011 | 1111001 |

Reed–Solomon coded block

able to detect and correct complete errored blocks. It is thus possible for a whole block to be corrupted owing to a burst of errors in a deep fade, for example, and the receiver still to be able to reinstate the correct information.

RS codes are often used in mobile radio systems where burst errors are common, either as an alternative to, or in addition to, interleaving. RS codes are also used as part of the error correcting mechanism in CD players. The inevitable scratches on the disk surface result in bursts of errors coming from the disc, making the RS code ideally suited to this task.

7.5 Convolutional coding

Introduction

In contrast to block codes, a *convolutional code* is implemented on a bit-by-bit (serial) basis for the incoming source data stream. The encoder has memory and executes an algorithm using a predefined number of the most recent bits to yield the new coded output sequence.

1001011010001 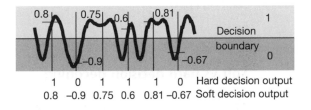 1011010000100011010

Convolutional coding

The decoding process is also usually a serial process based on present and previous received data bits (or symbols). Both the encoder and decoder can be implemented using recursive engines, with one of the most efficient and well known being the *Viterbi convolutional decoder*, named after Andrew Viterbi, the inventor (Viterbi, 1967).

Soft decision decoding

So far in this chapter, it has been implied that the decoding process is performed on the demodulated bit stream at the receiver output – the so-called *hard decision* data. A hard decision means that the decoder is given no information on how close the received symbol was to the decision boundary, before the decision was taken.

0.8	0.75	0.6	0.81		1
				Decision	
				boundary	
	–0.9		–0.67		0

1	0	1	1	1	0	Hard decision output
0.8	–0.9	0.75	0.6	0.81	–0.67	Soft decision output

By providing a decoder with 'side' information about how close each symbol was to the decision boundary the decoder is able to place greater or lesser emphasis on the validity of any given symbol in its decoding process. In essence, the decoder tries to *weight* the decisions and form an overall view over many symbols as to the most likely transmitted symbol sequence. This enhanced decoding process is called *soft decision* decoding.

Trellis diagrams

Trellis diagram with soft decision side information

A frequently used pictorial representation of the convolutional encoding and decoding process is the *trellis diagram*. At each stage in the trellis, a number of valid symbols or bit patterns exist, and over time, the decoder builds up a picture of the path exhibited by the received data through the trellis.

With soft decision decoding, each element in the path can be weighted with a confidence factor using the 'side' information. The job of the decoder is, in effect, to check all possible paths through the trellis (not just the most apparent one) and add up all the weights over N sections, to see which path in fact gives the highest confidence rating. This path is then chosen as the correct set of received symbols and is decoded to give the output bit stream.

The Viterbi algorithm is very effective at performing this path search process and in recent years has been effectively fabricated on silicon. More information on trellis coding and the Viterbi algorithm can be found in Ziemer and Peterson (1992).

Soft vs hard decision decoding

As would be expected, a soft decision decoding scheme has a better chance of detecting and correcting errors than a hard decision scheme.

If there is unlimited and unquantized weighting information available to the decoder, then the soft decision decoder should provide up to 3 dB *coding gain* improvement over a hard decision decoder for the same code length. If the weighting information is quantized to only eight levels, the coding gain is reduced to approximately 2.75 dB.

It is common for soft decision decoders to use eight or fewer weighting levels to avoid undue complexity in the decoder hardware/software.

7.6 Combined coding and modulation

Introduction

The problem with all the coding schemes described thus far is that they all increase the amount of data to be transmitted, requiring either increased transmission bandwidth or reduced channel capacity if the bandwidth is fixed.

There is, however, one obvious method to recover any capacity loss and that is to increase the number of symbol states in the modulation scheme. We know that this will incur a large E_b/N_0 penalty, but is this more or less than the coding gain that we can achieve with the extra channel capacity obtained?

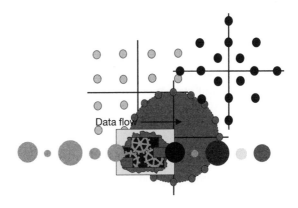

Data flow

Trellis Coded Modulation (TCM)

In 1976, G. Ungerboeck and I. Csajka published a very famous paper on this exact topic (Ungerboeck and Csajka, 1976), where they showed that in fact the potential coding gain could indeed outweigh the E_b/N_0 penalty of increased symbol states, by 3–4 dB! In a later paper (Ungerboeck, 1982) examples of M-ary PSK and M-ary QAM schemes with coding gains in excess of 6 dB were presented.

The efficient implementation of these coding schemes often makes use of convolutional encoding and the trellis-based Viterbi decoding algorithms mentioned earlier. For this reason, this type of coding is often referred to as *Trellis Coded Modulation (TCM)*.

Practical issues with TCM

TCM is now used extensively in modern high-speed telephone and radio modems. For example, a 28.8 kbps dial-up modem (V.34 standard) can use up to 1408-QAM, with a 16-state, rate 2/3 code.

There are some practical issues to note with TCM:

- Increasing the number of symbol states may incur an implementation penalty as well as the theoretical E_b/N_0 penalty, requiring greater amplitude and phase accuracy in the TX/RX system.

- More symbol states may result in an increased peak to mean ratio for the transmit hardware to cope with.

- When TCM systems are operated in very poor signal conditions, they 'fall over' much more rapidly than the uncoded equivalent throughput system.

TCM and the Shannon limit

The significant coding gain of TCM-based modems means that it is now possible to get very close to the Shannon limit for certain modulation types and coding levels. The implementation complexity may once have appeared daunting and almost impractical back in 1976 when the technique was first published (Ungerboeck and Csajka, 1976), but today TCM modems are fully integrated into a single IC and available as credit card (PCMCIA) sized plug-ins for PCs.

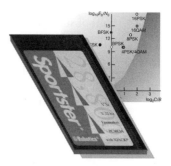

Source, channel and modulation coding – complete

Having highlighted the major benefits of combining channel (bit level) and modulation (symbol level) coding, it begs the question whether adding intelligent source coding to the pot will yield even further benefits.

The answer is undoubtedly yes, with some schemes already in use. Modern speech coders, for example, assign priority to certain bit patterns which are in turn assigned extra protection within the channel code. Similarly, key bits of information in an image are being given priority in the more recent image transmission systems.

A number of research groups are also looking at the viability of mapping the source information directly to M-ary symbols, bypassing the bit encoding level altogether, in an attempt to further integrate and optimize the source, channel and modulation processing tasks.

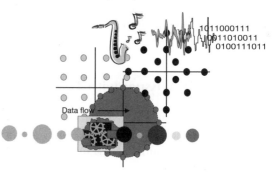

QUESTIONS

7.1 A PCM system uses a sampling rate of 8000 samples per second. What is the maximum input signal frequency that can be supported by this system without aliasing occurring? What is the minimum input frequency that can be supported?

7.2 What will be the output from A/D and D/A converters placed back to back, for an input signal consisting of two tones, one at 5000 Hz and one at 9000 Hz? The sampling rate for the system is 12 000 samples/second.

7.3 What A/D converter range is required to support a source signal having a dynamic range of up to 120 dB?

7.4 What will be the peak signal to quantization noise ratio for a 16-bit converter, expressed in dB?

7.5 Determine the parity check bits for the following data words, assuming *odd* parity:

(a) 101100111

(b) 0010100

(c) 0111011110

7.6 A data word with *even* single bit parity is received with three bit errors. Will the parity check process detect these errors?

7.7 A data stream with a bit rate of 12 kbps is coded using a block code prior to transmission which results in a coded bit rate of 20 kbps. What is the code rate being used for this system, and what is the redundancy of this particular block code?

7.8 A (127, 120) block code is used to provide error detection on a satellite link. If the coded on-air data rate is 254 kbps, what is the information transfer rate for the link?

7.9 A block code is designed with a Hamming distance of 5. How many errors can the code detect, and how many errors could it correct?

7.10 What Hamming distance is required for a block code that must correct for up to four errors in each block?

8 Multi-user digital modulation techniques

8.1 Introduction

For the majority of data communications that take place, there is a requirement for several users to *share a common channel* resource at the same time. This resource could be the high speed optical fibre links between continents, the frequency spectrum in a cellular telephone system, or the twisted pair 'ethernet' cable in the office.

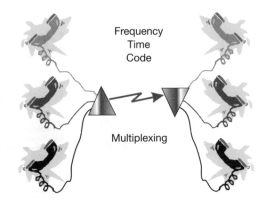

For multiple users to be able to share a common resource in a managed and effective way requires some form of *access protocol* that defines when or how the sharing is to take place and the means by which messages from individual users are to be identified upon receipt. This sharing process has come to be known as *multiplexing* in wired communication systems, and *multiple access* in wireless digital communications.

Three classes of multi-user access techniques will be considered in this chapter: techniques where individual users are identified by assigning different *frequency* slots, techniques where users are given different *time* slots, and techniques where users are given the same time and general frequency slots, and are identified by different *codes*. We have already touched upon time-based multiplexing techniques when looking at packet transmission in the section on the fundamentals of data networks and protocols in Chapter 1.

8.2 Frequency Division Multiple Access (FDMA)

Basic system operation

Used extensively in the early telephone and wireless multi-user communication systems, *frequency division multiplexing* of users is perhaps the most intuitive form of resource sharing.

If a channel, such as a cable, has a transmission bandwidth W Hz, and individual users require B Hz to achieve their required information rate, then the channel in theory should be able to support W/B users simultaneously by using bandpass modulation, and placing each user in an adjacent slot of the available bandwidth. Immediately, we see that the efficiency of frequency multiplexing is

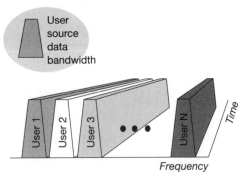

Frequency division multiple access

governed by how effectively the transmission bandwidth is constrained by each user (the value of α in the root raised cosine filter, for example). It is also dependent on how good (selective) the 'de-multiplexing' system is at filtering out the modulation corresponding to each user.

With frequency division multiplexing, the data rate and hence modem design for each user remains unchanged by the requirement to operate a multi-user system, and the only additional circuitry is for frequency conversion to the assigned slot. The user will typically be assigned the frequency slot for the duration of the message.

Wireless FDMA operation

FDMA is widely used in wireless communications systems where the radio environment creates several challenges for any multiple access method owing to the unpredictable and time varying nature of the communications channel.

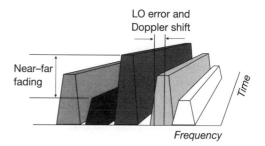

Frequency division multiple access

One of the biggest challenges is the very large variations in received signal power that arise from users in different frequency slots due to the so-called *near–far* effect. A radio user that is very near to a base-station receiver will produce a much stronger signal than that from a distant (far) user operating on the extreme of the communication range. Typical variations in power can be up to 100 dB. If the strong signal is producing any out-of-band radiation in the slot occupied by the weak signal, this can easily swamp the weak signal, corrupting the communications. Much of the discussion in this book on controlling the bandwidth and side-lobe energy of digital modulation formats, such as CPFSK (see Section 5.3), and on designing modulation formats that are not overly sensitive to amplifier distortion, such as $\pi/4$ QPSK (see Section 6.4), are all driven by this near–far problem in the wireless application.

Other challenges in the radio environment include dealing with the frequency uncertainty for any individual user caused by Doppler shift and local oscillator error (see Section 4.2). This inevitable error requires *guard-bands* to be allocated between frequency slots, thus sacrificing some of the efficiency of the FDMA scheme.

Power control in wireless FDMA systems

As the near–far problem can be so dominant in wireless FDMA operation, it is worth looking briefly at techniques for alleviating the problem. The most

effective technique over and above
maximizing the filtering and spectral shaping
within the modem and improving linearity in
the TX/RX subsystems is to attempt to level
out the signal power from each user at the
receiver site. If each user were able to fully
control the output power from its own
transmitter, and assuming that it knew the
path loss to the receiver, then it could adjust
its own power to ensure a fixed minimum
(yet sufficient) level from all users at the
receiving site.

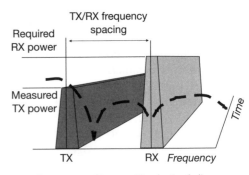

Power control issues with selective fading

Determining the path loss is the key problem. Certainly it is possible in a
duplex link for the remote user to measure the power received from the base-
station site and hence calculate the path loss in the 'downlink'; however,
unless the user is stationary and operating on the same frequency for transmit
and receive, this does not necessarily translate to the path loss in the 'uplink'
direction. For example, the remote user could be a receiver in a frequency
selective fade, in which case the unit would overestimate the path loss
involved. Alternatively the uplink could be subject to a frequency selective fade
and the unit not generate sufficient transmit power. A solution to this problem
is to operate a *closed loop power control* system whereby the base-station unit
monitors the signal power from each remote unit and issues commands to
increase or decrease TX power accordingly. This can, however, incur a high
signalling overhead in the system. It turns out that power control is very
critical in CDMA and to a lesser extent TDMA operation, and closed loop
power control is common in CDMA applications.

Advantages of FDMA

Traditionally FDMA has been favoured
for use in radio systems, where the path
delays introduced by multipath
propagation give rise to intersymbol
interference effects (see Section 3.2)
which become significant when the
differential path delay becomes a
significant part of the symbol period.
By keeping the symbol duration high,
which implies M-ary signalling over narrow frequency slots, the delay problem
can often be ignored. (With modern signal processing, the implementation of
channel equalization techniques (see Section 4.5) has meant that this delay
spread problem need no longer constrain the symbol rate used, allowing much

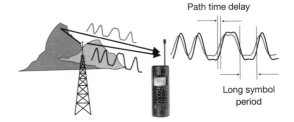

Path time delay

Long symbol
period

faster data rates over air and hence opening up the possibility of TDMA solutions as an alternative to FDMA.)

Another advantage of FDMA is that the bandwidth of the TX and RX circuitry is kept to a minimum (particularly the bandwidth over which power amplifiers are to be made linear), and the signal processing task for data generation and detection is kept as simple as possible.

Examples of FDMA use include the first-generation cellular telephones and the majority of two-way radio systems in use by taxi companies, trucking fleets, emergency services and so on.

Disdvantage of FDMA

A disadvantage that has often been levelled at FDM/FDMA is the inflexibility to accommodate variable user data rates within a fixed bandwidth frequency slot. This claim is nowadays unfounded for two reasons. Firstly, it is practical to vary the data rate in a given frequency slot by increasing the number of symbol states used. Secondly, it is possible to assign a user more than one frequency slot, or introduce the concept of a variable bandwidth slot in order to vary the user data rate. Both of these solutions rely heavily on the advent of powerful digital signal processing devices that can implement adaptive rate multi-symbol modems (these are now commonplace in line modem cards) and variable bandwidth matched channel filters – again a simple function for today's DSP devices.

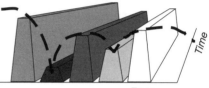

Frequency selective fading

Frequency stability and the need for guard-bands has traditionally been a bigger problem for FDMA use, requiring very costly and high stability oscillators in the modems if the guard-bands are to be kept to a minimum. In recent years, the use of a broadcast 'off-air' reference has been exploited to allow designers to dispense with these costly oscillator components and achieve much greater stability than hitherto possible. (Today, it is possible to buy watches that take their timing reference 'off-air' for millisecond precision accuracy.) The major disadvantage of FDMA in a wireless environment is the susceptibility of any individual narrow frequency slot to frequency selective fading (see Section 4.5) which can cause loss of signal for that user – usually on a temporary basis.

Wavelength division multiplexing

In optical fibre communications, it has been very difficult until recently to generate and detect individual carrier frequencies of light with sufficient

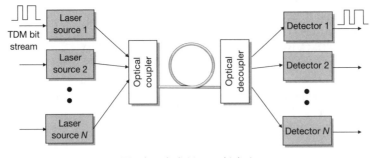

Wavelength division multiplexing

spectral stability to be sandwiched within the finite bandwidth available on long lengths of fibre and still resolved at the receiver. Recently, however, the laser source, repeater amplifier and detector technology have all improved to permit several independently modulated light carriers to be sent over a single fibre. In optical circles, this is referred to as *Wavelength Division Multiplexing* (*WDM*) rather than frequency division multiplexing.

Each individual light carrier would typically be supporting data rates of up to 10 Gbps with users 'time multiplexed' onto the channel. WDM thus offers the possibility of several hundreds of gigabits transmission over a single fibre and also bi-direction transmission over the same fibre.

8.3 Time Division Multiple Access (TDMA)

Basic system operation

The basic principle behind time division multiplexing is that the user has access to a modem operating at a rate *several times* that required to support his own data throughput, such that he can send his information in a time slot that is shorter than his own message transaction. Other users can then be assigned similar time slots on the same channel. Clearly if the data rate on the channel is w bits/second, and each individual user requires only b bits/second, then the system can support w/b simultaneous users.

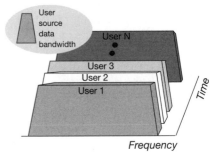

Time division multiple access

In many TDM systems, users are assigned a time slot for the duration of their call whether they require it or not. So, for example, if the user is generating voice traffic, or typing at a keyboard, a time slot will be assigned regularly regardless of whether the person is speaking or a key has been pressed, and it is very likely that the channel capacity is being 'wasted' (this is equally true of FDMA systems).

In order to maximize the use of a channel resource under these circumstances, packet-based transmission (see Section 1.3) is now common on wired links, where the user is not given a fixed repeated time slot, but rather allocated a time slot 'on demand'. This system works well provided time slot availability can be guaranteed for real-time applications – video, voice and so on. It also involves quite a high penalty in signalling overheads.

TDMA in a wireless environment

Just as for FDMA, the wireless environment provides particular challenges to TDMA operation. Again, the 'near–far' effect comes into play, with signals from a distant user taking longer to arrive at the base-station than those from a near user. In order to accommodate these delays, *guard-times* are required between time slots (cf. frequency guard-bands in FDMA) both to accommodate

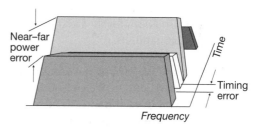

Time division multiple access in a wireless environment

the variable delay between near and far users, and also to allow for timing errors in the start of a time slot transmission by any individual user.

The near–far problem also gives rise to the same signal strength fluctuations in the base-station receiver as identified for FDMA, but in this case there is no problem with adjacent channel interference (see Section 4.3) as no user is operating concurrently with another. The receiver is, however, required to react very rapidly to the changing power level from users in different time slots, and power control (see Section 8.2) of each user is commonly applied to alleviate this problem.

Example of a TDMA system

The GSM digital cellular system is a very good example of a TDMA-based *air interface* that has been designed to cope with the challenges of the wireless environment.

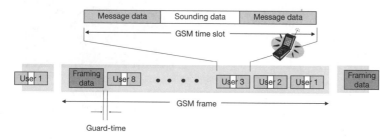

GSM TDMA data and frame structure

We have already seen that GSM incorporates a reference word within each frame for channel equalization (see Section 4.5), necessary to overcome the multipath delay problems which FDMA seeks to avoid. The system also involves a technique known as *time slot advance* where the remote unit measures the time delay for information to be sent on the *downlink* from base-station to mobile, and then automatically advances the start time of its own *uplink* transmission in order to compensate for the uplink time delay. This technique, together with each mobile taking a master timing reference from the base-station, allows the guard-times between TDMA slots to be minimized and also alleviates the need for an accurate and costly timing reference in the cellular handsets.

The GSM frame format is such that eight users are assigned to the same transmission frequency, and hence eight time slots are provided in each frame, to be repeated regularly in subsequent frames. Each user must transmit information at a rate of 270 kbps within the 200 kHz bandwidth allowed (GMSK is used), even though the individual user data rate from the voice coder is only 13 kbps.

The throughput of each modem is thus much higher than required to support the users' data, and hence one could expect the design to be much more complicated and costly than an FDMA equivalent. This need not be the case, however, if proper advantage is taken of the fact that the data can be stored and processed 'at leisure' over the seven remaining time slots as well as during the arrival time slot.

Adjacent GMSK spectrum for GSM

Advantages of TDMA operation

An often quoted advantage of TDMA operation is the ease with which users can be given variable data rate services by simply assigning them multiple time slots. For example, a GSM user could be given all eight time slots within a frame, giving a total fully coded user data rate of $8 \times 9600\,\text{bps} = 76\,800\,\text{bps}$. (This assumes that the user's equipment has sufficient processing power to process all eight time slots simultaneously.) This facility has its dual in assigning multiple frequency slots

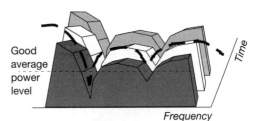

TDMA in Frequency selective fading

in an FDMA system; however, this is perceived to require more complex hardware for implementation.

A second advantage of TDMA is the commonality in the base-station of the transmitting hardware for all time slot users. There is only one power amplifier required to support multiple users (albeit with a wider modulation bandwidth). Traditionally with FDMA, each user channel at the base-station has required an individual power amplifier, the output of which is combined at high power to feed a single common antenna. The advent of multicarrier linear power amplifiers (see page 83) in the past few years has begun to alter this TDMA/FDMA bias.

For packet-based applications, TDM/TDMA operation is clearly a well-matched access method.

Disadvantages of TDMA operation

One of the more challenging aspects of TDM/TDMA operation is the establishment of *system timing* in order to ensure correct time slot arrival and framing and to cope with variable path delays in the wireless systems. While these can be accommodated by careful design, for small multi-user communication systems, the overhead of system timing may favour the use of FDMA methods.

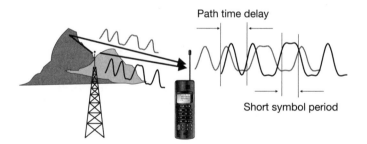

Path time delay

Short symbol period

TDMA use also requires each user terminal to support a much higher data rate than the user information rate. This implies faster processing for modulation and demodulation, wider bandwidth in the TX/RX section and in particular requires a higher peak power rating for the power amplifier in a wireless application compared with an FDMA solution. (Assuming equal symbol average energy for both systems to achieve equivalent range, the symbol duration for the TDMA system must be much shorter than the FDMA equivalent and hence the transmitted symbol power for TDMA must be correspondingly higher than for FDMA. Note: the average symbol power and hence average battery drain for both systems are identical.)

8.4 Code Division Multiple Access (CDMA)

CDMA systems

Traditionally *Code Division Multiple Access (CDMA)* systems have been used almost exclusively by the military as a means of operating covert radio communications in the presence of high levels of interference. In recent years, the interference immunity of CDMA for multi-user communications, together with its very good *spectral efficiency* characteristics, has been seen to offer distinct advantages for public cellular-type communications.

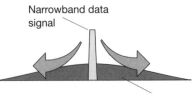

Narrowband data signal

Spread signal

There are two very distinct types of CDMA system, classified as *direct sequence* CDMA and *frequency hopping* CDMA. Both of these systems involve transmission bandwidths that are many times that required by an individual user, with the energy of each user's signal *spread* with time throughout this wide channel. Consequently these techniques are often referred to as *spread spectrum* systems. One of the most prominent proponents of CDMA in the cellular market is Andrew Viterbi, and his book on the subject (Viterbi, 1997) is well worth reading.

Frequency hopped CDMA (FH-CDMA)

Frequency hopping involves taking the narrow bandpass signals for individual users and constantly changing their positions in frequency with time. In a frequency selective fading environment, the benefit of changing frequency like this is to ensure that any one user's signal will not remain within a fade for any prolonged period of time. Clearly for frequency hopping to be

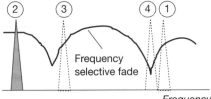

Frequency selective fade

Frequency

effective, the users must hop over a bandwidth significantly wider than *notch* caused by frequency selective fading. In order to ensure that individual users never (or rarely) hop onto the same frequency slot at the same time, causing mutual interference, the carrier frequencies are assigned according to a predetermined sequence or *code*.

Frequency hopping is most effective if a fast hopping rate is used (several thousand times per second) so that the communications are not corrupted by fading or mutual interference for any length of time. This, however, brings problems in the design of fast switching synthesizers and broadband power amplifiers which in practice put an upper limit on the hopping rate. Also, the narrowband channels are susceptible to Doppler shift, local oscillator error and so on (see Section 4.2), and compensation techniques are struggling with fast hop rates. Hopping also makes a system less vulnerable to discrete narrowband interference and near–far effect problems.

Example of frequency hopped CDMA

We have already classed GSM European Digital Cellular as a TDMA system, but it also has provision within the standard to change frequency on a frame-by-frame basis, making it a modest rate (Y hops/second) frequency hopping CDMA system.

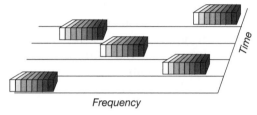

Frequency hopping with GSM

The motivation for adding the extra complexity of hopping to GSM is twofold. Firstly, the 200 kHz channel bandwidth of GSM is not sufficient to ensure that it will always be significantly wider than the coherence bandwidth of the multipath environment, and thus not corrupted by narrowband fading. Secondly, if there is a strong interference source on any given channel, the hopping process will ensure that frames are only corrupted on an occasional basis.

A further example of an FH-CDMA system is the GTI Geonet radio, a high capacity public and private two-way radio solution. This product uses 25 kHz spaced channels transmitting at 36.9 kbps with $\pi/4$ QPSK modulation. The hop rate is faster than GSM at 152 hops/second, with a TDMA slot width of 2.2 ms and three slots per frame.

Direct sequence CDMA

In *direct sequence CDMA*, the narrowband signals from individual users are spread continuously and thinly over a wide bandwidth using a *spreading*

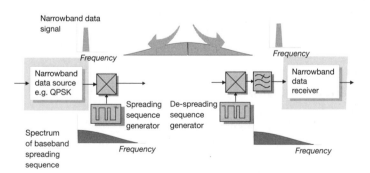

Direct sequence CDMA implementation

sequence. By mixing the narrowband user data signal with a locally generated well-defined wideband signal, the user energy is spread to occupy roughly the same bandwidth as the wideband source. The wideband spreading signal is generated using a *pseudo-random sequence generator* (see in-depth section) clocked at a very high rate (termed the *chipping rate*).

De-spreading of the signal is necessary in the receiver in order to recover the narrowband user data modulation and this is accomplished by mixing the received signal with an identical, accurately timed pseudo-random sequence. This correlation process has the effect of reversing the spreading action in the transmitter. De-spreading will only occur, however, if the correct sequence is used at both ends of the link, and if the two sequences are time aligned (see in-depth section).

Multi-user operation is achieved in direct sequence CDMA by assigning each user a different spreading code or a different time alignment of a common spreading code. Only that portion of the wideband spectral energy that has been spread by the same code as used in the receiver will be detected. Users are thus able to coexist in the same bandwidth and time space on the channel.

Like frequency hopping, spread spectrum CDMA overcomes the problem of frequency selective fading by ensuring that most of the spread signal energy falls outside the fading 'notches'.

Code Division Multiple Access
(Direct Sequence)

If there is some correlation between spreading codes, as is almost always the case, then there will be a small contribution to any individual de-spread user signal from all the other spread users on the channel. Ultimately this puts an upper limit on the number of users that can co-locate on the same piece of spectrum before the unwanted de-spread energy gives rise to unacceptable data errors. This interference factor also gives rise to one of the most stringent power control requirements of all access techniques, as it is clear that this mutual interference between users will be minimized for each user if they all operate to give an identical spread power level at the receiving site.

Example of direct sequence CDMA

The most widely adopted CDMA radio system (outside the military) was pioneered by QUALCOM, a California-based company, and is now embodied in the IS-95 standard for cellular telephone applications. This 'air interface' is a direct sequence CDMA design, spreading each user voice or data signal over a 1.25 MHz channel bandwidth.

IN DEPTH

Pseudo-random sequences

One of the most common ways of spreading a CDMA signal is with a pseudo-random noise sequence (often called a PN sequence). This is a sequence generated by a shift register with feedback (see diagram) which repeats itself after every $N = 2^n - 1$ clock cycles. The 3-bit shift register shown would thus generate a sequence that repeats every seven bits (often called 'chips' in spread spectrum applications).

Of particular interest in CDMA is the auto-correlation of a PN sequence with itself, and the cross-correlation with other PN sequences generated using different length shift registers, or registers with different 'taps' from which the feedback signals are obtained.

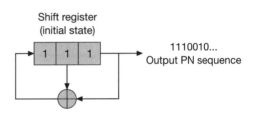

Simple 3-stage PN sequence generator

The result of *auto-correlation* of a PN sequence is shown here in the figure. Auto-correlation in this application involves comparing the similarity of one sequence with a time-displaced version of the same sequence for all possible time offsets until the sequence repeats itself. A maximum output is achieved only when the two sequences are exactly time aligned, and falls to a minimum value of $1/N$ for all sequence offsets greater than 1 chip period. Auto-correlation is achieved practically by mixing the incoming spread spectrum signal with a locally generated replica of the spreading PN sequence and sliding the timing of the local PN sequence until a correlation peak is reached. At this point, the modulation, which has been superimposed on the spreading code in the transmitter, can be extracted.

If all other spread spectrum signals are operating with unique and carefully chosen spreading codes, then there will be no *cross-correlation* between them and the contribution from all these other users will also be $1/N$ times that of the wanted user correlation peak. Clearly, the larger the sequence length N, the larger the wanted correlation peak will be with respect to all other signals. This is very important if a large number of users are to be accommodated on the same frequency while still allowing the de-spreading of any individual signal with sufficient signal-to-noise ratio for

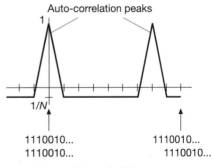

Auto-correlation of a PN sequence

acceptable communication quality. If the sequence length N is made too big, however, the time taken to find the correlation peak by sliding one sequence against another can be prohibitively long.

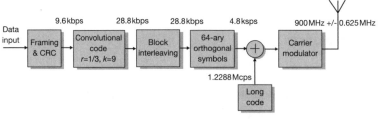

Uplink modulator for IS-95 CDMA

The maximum user data rate per spreading code is 9600 bps (rates of 1200, 2400 and 4800 are also possible), which is then channel coded up to 19 200 bps in the forward channel (base to mobile), and 28 800 bps in the reverse channel (mobile to base).

In the reverse channel, the 28.8 kbps coded data is mapped onto 64 orthogonal symbol states known as *WALSH functions*, giving a symbol rate of 4800 symbols/second. This narrowband symbol stream is then spread using a unique identifying spreading code for every mobile unit – called the long code, clocking at a rate of 1.2288 million chips/second. The long code is a pseudo-random sequence of length $2^{24} - 1$ bits. This means that there are $2^{24} - 1 = 4\,398\,046\,511\,103$ possible different code sequences, enough for all the mobiles in the world!

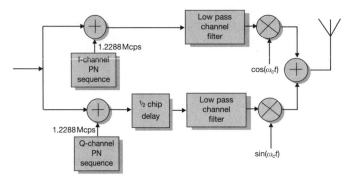

Offset QPSK modulator for IS-95 CDMA

The spread baseband sequence in IS-95 must then be modulated onto the carrier frequency for transmission. In both the forward and reverse link channels, two further second short code PN sequences are superimposed onto identical versions of the spread baseband stream, which are then filtered to restrict the signal bandwidth. The specification calls for a stop band rejection of better than 40 dB at 740 kHz.

These two filtered spread baseband data streams form inputs to a *QPSK modulator* (see Section 6.4). In the forward path this is a standard QPSK approach; however, in the reverse link, *Offset QPSK* is used (see Section 6.4) making full use of the non-zero envelope properties of this technique to alleviate the design of the handset RF power amplifiers.

In addition to the main traffic channel characteristics described, there are further pilot, synchronization and paging channels in the IS-95 standard for system control in this highly complex digital communications link.

Advantages of CDMA

Spreading the user signal well beyond the frequency selective fading bandwidth is clearly advantageous for coping with this unique problem in wireless communications. It also provides protection from narrowband interfering signals (these are spread by the 'de-spreading' process in a direct sequence CDMA receiver).

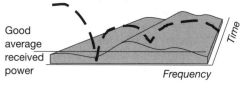

CDMA in selective fading

Perhaps the main advantage of CDMA as a multi-user system is the *flexibility to accommodate variable user data capacity*. Each user in a spread spectrum CDMA system can increase their modulation rate and local narrowband modulation bandwidth without affecting other users when it is spread, as long as the user does not increase the overall wideband energy of the composite multi-user signal and hence increase the chance of mutual interference when de-spreading beyond that tolerable to the system.

By slightly over-subscribing the number of users and their 'spread energy quota' on a spread spectrum CDMA system, it is possible for an operator to exploit the fact that any individual user will not be using the 'corporate' channel all of the time, and provided that the transmitter is powered down when the user is not speaking, for example, the average overall energy will be at the acceptable 'under-subscribed' level.

Disadvantages of CDMA

Wideband spreading of signals, whether it be with 'frequency hopping' or 'direct sequence' techniques, has a penalty in terms of the signal processing overhead involved with such high rate and bandwidth transmission. Power control has also been identified as a critical issue in maximizing the number of users that can be supported on a given common frequency channel.

CDMA also requires a large amount of bandwidth to be available in a contiguous block (spread spectrum only) in order to ensure that sufficient spreading can be obtained to mitigate the frequency selective fading and to ensure that there is sufficient coding gain in the system. Typically bandwidths of 5 MHz upwards are desirable for best communications performance in a typical cellular environment, although regulatory constraints have forced smaller bandwidths (1.25 MHz for IS-95) to be used in some circumstances.

8.5 Combined multiple access systems

Examples of FDMA/TDMA combinations

We have already seen some examples of digital communication systems exploiting combinations of multi-user access techniques. GSM, although primarily a TDMA system, requires several 200 kHz frequency channels (each carrying eight time slots) in order to provide a practical high

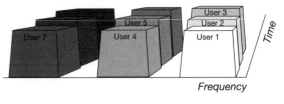

FDMA/TDMA combinations

capacity cellular system and can thus be viewed as an FDMA system also.

The TETRA (Trans European Trunked Radio Access) system exploits a three time slot TDMA structure with FDMA channels spaced at 25 kHz spacing.

Examples of FDMA/CDMA and FDMA/FDMA combinations

An example of a CDMA/FDMA combination is the IS-95 cellular system described earlier which uses spread spectrum over 1.25 MHz channels, with a number of these wideband channels being used to make up a typical cellular service.

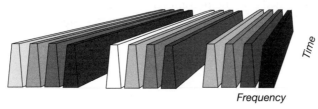

FDMA/FDMA operation

The GSM system we have seen is both a TDMA and frequency hopped CDMA technique.

There are even some applications of so-called FDMA/FDMA systems such as the DC/MA system (see in-depth section), where 25 kHz FDMA channels are further subdivided by frequency into five or more individual voice and data channels, each user being able to access some or all of these channels to give a higher composite data rate.

Dynamic Channel Multicarrier Access (DC/MA)

DC/MA is a proprietary two-way radio air-interface designed to be a direct upgrade for 25 kHz channelled FM analogue radio systems. The DC/MA approach is a further FDMA subdivision of the 25 kHz channel to give five voice or data channels within the same bandwidth as the original single FM voice channel.

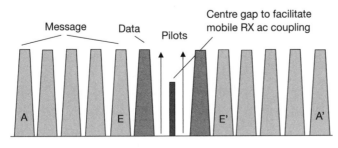

Downlink DC/MA channelization

The downlink channel structure incorporates dual pilot tones for channel sounding and frequency/Doppler correction, and an additional dedicated data channel for system trunking and so on. Each of the downlink channels is split into two sub-channels placed symmetrically about the channel centre – this has been termed the 'onion ring' approach. In data mode, each sub-channel pair can support 9600 bps data using a 16-QAM modulation format.

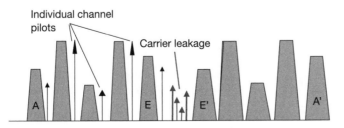

Uplink DCMA channelization

The uplink channel structure differs from the downlink structure in that each uplink channel requires its own pilot tone so that the receiving base-station has an independent measure of the fading and frequency error from each mobile user. The additional bandwidth required to support the five pilot signals means that the uplink cannot support the sixth dedicated data channel found in the downlink.

Glossary

A/D conversion (analogue-to-digital conversion)   The process of converting a time-continuous analogue signal to a sampled digital representation

Additive White Gaussian Noise (AWGN)   The term describing a noise signal which has a flat power spectral density with frequency

Amplitude Shift Keying (ASK)   A digital modulation format where information is conveyed in the amplitude of a carrier signal

Automatic Repeat Request (ARQ)   Protocol for dealing with data words that are corrupted by errors whereby the receiving system requests a re-transmission of the word(s) in error

BCH   A type of block coding of data words named after the inventors Bose, Chaudhuri and Hocquenghem

Binary Phase Shift Keying (BPSK)   Digital modulation format where information is conveyed in the phase of a carrier signal: usually $0°$ and $180°$

Bit error rate (BER)   A measure of the number of bit errors occurring in a transmitted data sequence

Code Division Multiple Access (CDMA)   Method of combining multiple users on a given channel bandwidth using unique spreading codes, or hopping patterns to distinguish any given user

Continuous Phase Frequency Shift Keying (CPFSK)   Digital modulation format where information is conveyed in the frequency of a carrier signal ensuring that the phase is continuous between symbol transitions

D/A conversion (digital-to-analogue conversion)   The process of converting a discrete time digital representation of a waveform to a continuous analogue voltage equivalent

dBc   The level of a signal in dB relative to a wanted carrier signal level

dBm   the level of a signal in dB relative to $1\,mW$ ($0\,dBm$)

Differential Phase Shift Keying (DPSK)   Digital modulation format where information is conveyed in the phase difference of a carrier signal between consecutive symbols

Differentially Encoded Phase Shift Keying (DEPSK)   Digital modulation format where data is pre-coded to convey information in the change of state between consecutive bits, to overcome the phase ambiguity problem with coherent phase shift keying

Digital Audio Broadcasting (DAB)   The generic name given to the new generation of radio transmission using digitally encoded audio waveforms

Digital Video Broadcasting (DVB)   The generic name for the new generation of image and sound transmission using a digitised version of the image signal

Dynamic Channel Multi-carrier Architecture (DC/MA)   A proprietary radio system designed by Wireless Systems International for ComSpace Corporation

E_b Symbol denoting the energy required to represent 1 bit of information in the modulated signal

E_s Symbol denoting the energy required to represent 1 symbol of information in the modulated signal

European Telecommunications Standards Institute (ETSI)   The body in Europe responsible for setting standards relating to wireless communications

Feed Forward Signal Regeneration (FFSR)   A method for correcting for multipath fading and frequency error in mobile radio systems using a pilot reference tone

Forward Error Correction (FEC)   An error correction method allowing detection and correction of bit or word errors without the need to re-transmit the data

Frequency Division Multiple Access (FDMA)   Method of combining multiple users on a given channel bandwidth using unique frequency segments

Frequency Shift Keying (FSK)   Digital modulation format where information is conveyed in the frequency of a carrier signal

Gaussian Minimum Shift Keying (GMSK)   Digital modulation format where information is conveyed in the frequency of a carrier signal, where the incoming data is first shaped with a Gaussian low-pass filter

GSM (Global System for Mobile communications)   A cellular radio standard using digital GMSK modulation

Intermediate Frequency (IF)   A special frequency used in radio systems as part of the up-conversion or down-conversion process; usually where the channel selection is made

Intersymbol interference (ISI)   Interference between adjacent symbols often caused by system filtering, dispersion in optical fibres, or multipath propagation in radio system

International Telecommunications Union (ITU)   The body overseeing standards and regulation for the wired communications industry

M-ary Amplitude and Phase Keying (M-ary APK)   Digital modulation format where information is conveyed in the amplitude and phase of a carrier signal

M-ary Amplitude Shift Keying (M-ASK)   Digital modulation format where information is conveyed in the 'M' amplitudes of a carrier signal

M-ary Phase Shift Keying (M-PSK)   Digital modulation format where information is conveyed in the 'M' phases of a carrier signal

M-ary Quadrature Amplitude Modulation (M-QAM)   Digital modulation format where information is conveyed in the 'M' amplitude and phase combinations of a carrier signal

Minimum Shift Keying (MSK) Digital modulation format where information is conveyed in the frequency of a carrier signal with modulation index equal to half the symbol period

N_0 The noise power density measured in Watts/Hz

Non Return to Zero (NRZ) Binary data encoding format

ON-OFF Keying (OOK) Digital modulation format where information is conveyed by switching on and off a carrier signal

Orthogonal Frequency Division Multiplexing (OFDM) Modulation format where information is sent over multiple parallel adjacent frequency channels where the carrier frequencies are orthogonal

P_b Probability of receiving a data bit in error

PCS (Personal Communications System) Common term for cellular radio predominantly in North America

Phase Locked Loop (PLL) Feedback control circuit for tracking the frequency and phase of an incoming signal, often used in modem carrier recovery circuits

P_s Probability of receiving a data symbol in error

Quadrature Amplitude Modulation (QAM) Digital modulation format where information is conveyed in the amplitude and phase of a carrier signal

Quadrature Phase Shift Keying (QPSK) Digital modulation format where information is conveyed in four equi-spaced phases of a carrier signal

RF Radio Frequency

rms root mean square

RS (Reed–Soloman) A type of block code for error detection and correction

RX Short for receiver

S/N ratio Signal-to-noise ratio

Symbol Error Eate (SER) The probability of receiving a symbol in error (not to be confused with Bit Error Rate)

T_b The period of 1 data bit

TETRA (Trans European Trunked Radio Access) The name of the ETSI-standardised radio system for private and public mobile two-way radio applications

Time Division Multiple Access (TDMA) Method of combining multiple users on a given channel bandwidth using unique time segments

Trellis Coded Modulation (TCM) Method of coding multiple symbols to give improved symbol error rate performance in noisy conditions

TX Short for transmitter

Voltage Controlled Crystal Oscillator (VCXO) A circuit whose oscillating frequency is tied to that of a crystal (fundamental or harmonic)

Voltage Controlled Oscillator (VCO) A circuit whose output frequency varies in sympathy with an input voltage

Wavelength Division Multiplexing (WDM) Method of combining multiple users on a given channel bandwidth using unique wavelengths (or frequencies as in FDMA)

References

Bateman A. (1989) A general analysis of bit error probability for reference-based BPSK mobile data transmission. *IEEE Trans.*, **COM-37**, April, 398–402

Bateman A. (1990) Feedforward transparent tone-in-band: its implementation and applications. *IEEE Transactions on Vehicular Technology*, **39**(3), August, 235–43

Bateman A. (1992) The Combined Analogue Loop Universal Modulator – CALLUM. *Proc. 42nd IEEE Vehicular Technology Conference*, Denver, USA, May, 759–63

Bateman A. and McGeehan J.P. (1983) Theoretical and experimental investigation of feedforward signal regeneration. *IEEE Transactions on Vehicular Technology*, **32**, 106–20

Bateman A. and Yates W. (1989) *Digital Signal Processing Design.* Computer Science Press

Carlson B.A.(1986) *Communication Systems.* New York: McGraw-Hill

Chan K.Y. and Bateman A. (1992) The performance of reference-based Mary PSK with Trellis Coded Modulation in Rayleigh fading. *IEEE Transactions on Vehicular Technology*, **41**(2), May, 190–8

Costas J.P. (1956) Synchronous communications. *Proc. IRE*, **44**, 1713–18

Gardner F.M. (1966) *Phaselock Techniques.* New York: Wiley

Halsall F. (1992) *Data Communications, Computer Networks and Open Systems.* Harlow: Addison Wesley Longman

Haykin S. (1989) *An Introduction to Analogue and Digital Communications.* Wiley

Hetzel S.A., Bateman A. and McGeehan J.P. (1991) LINC transmitters. *Electronic Letters*, May

Ifeachor E.C. and Jervis B.W. (1993) *Digital Signal Processing – A Practical Approach.* Harlow: Addison Wesley Longman

Jakes W.C. ed. (1993) *Microwave Mobile Communications.* IEEE Press. Re-issue

Lindsey W.C. and Simon M.K. (1972) *Synchronisation Systems in Communications.* Englewood Cliffs, NJ: Prentice-Hall

Lucky R.W., Saly J. and Welden S.J. (1968) *Principles of Data Communications.* New York: McGraw-Hill

Mansell A.R. and Bateman A. (1996) Adaptive predistortion with reduced feedback complexity. *Electronics Letters*, **32**(13), 1153–4

Papamichalis P.E. (1987) *Practical Approaches to Speech Coding.* Englewood Cliffs, NJ: Prentice-Hall

Parsons K.J. and Kenington P.B. (1994) The efficiency of a feedforward amplifier with delay loss. *IEEE Transactions on Vehicular Technology*, **43**(2), May, 407–12

Petrovic V. (1983) Reduction of spurious emission from radio transmitters by means of modulation feedback. *Proc. IEE Conference on Radio Spectrum Conservation Techniques*, pp. 44–9, September

Proakis J.G. (1989) *Digital Communications.* Singapore: McGraw-Hill

Proakis J.G. and Manolakis D.G. (1992) *Digital Signal Processing: Principles, Algorithms, and Applications.* USA: Macmillan Publishing Company

Shannon C.E. (1948a) A mathematical theory of communications. *Bell Systems Technical Journal*, **27**, July, 379–423

Shannon C.E. (1948b) A mathematical theory of communications. *Bell Systems Technical Journal*, **27**, October, 623–56

Schwartz M. (1990) *Information, Transmission, Modulation and Noise.* McGraw-Hill

Ungerboeck G. (1982) Channel coding with multi-level phase signals. *IEEE Trans. Information Theory*, **IT-28**, January, 55–67

Ungerboeck G. and Csajka I. (1976) On improving data-link performance by increasing the channel alphabet and introducing sequence coding. *Proc. Int. Conf. Infomation Theory*, Ronneby, Sweden

Viterbi A.J. (1967) Error bounds for convolutional codes and an asymptotic optimum decoding algorithm. *IEEE Trans. Information Theory*, **IT-13**, 260–9

Viterbi A.J. (1997) *CDMA – Principles of Spread Spectrum Communications.* Reading, MA: Addison Wesley Longman

Wilkinson R. and Bateman A. (1989) Linearisation of Class C amplifiers using Cartesian feedback. *Proc. IEEE 1989 Workshop on Mobile and Cordless Telephone Communications*, September, 62–6

Young P.H. (1991) *Electronic Communications Techniques.* USA: Maxwell-MacMillan

Ziemer R. and Peterson R. (1992) *Introduction to Digital Communications.* Macmillan

Index